VERY-LARGE-SCALE INTEGRATION

Edited by **Kim Ho Yeap** and **Humaira Nisar**

Very-Large-Scale Integration
http://dx.doi.org/10.5772/65525
Edited by Kim Ho Yeap and Humaira Nisar

Contributors

Kim Ho Yeap, Ian Andrew Grout, Muhaned Zaidi, Abu Khari A'Ain, Vadim Glebovsky, Kittipong Tripetch, Changyeol Lee, Vithyalakshmi N, Ashok Kumar N, Nagarajan P, Vinoth G.S

Published by InTech
Janeza Trdine 9, 51000 Rijeka, Croatia

Publishing Process Manager Edi Lipovic
Technical Editor SPi Global
Cover InTech Design team

First published February, 2018
Printed in Croatia
Legal deposit, Croatia: National and University Library in Zagreb

Additional hard copies can be obtained from orders@intechopen.com

Very-Large-Scale Integration, Edited by Kim Ho Yeap and Humaira Nisar
p. cm.
Print ISBN 978-953-51-3863-1
Online ISBN 978-953-51-3864-8

For EU product safety concerns:
IN TECH d.o.o., Prolaz Marije Krucifikse Kozulić 3, 51000 Rijeka, Croatia,
info@intechopen.com or visit our website at intechopen.com.

Meet the editors

Kim Ho Yeap is an Associate Professor at Universiti Tunku Abdul Rahman, Malaysia. He is an IEEE senior member and a chartered engineer registered with the UK Engineering Council. He received his MSc degree in microelectronics from Universiti Kebangsaan Malaysia in 2005 and his PhD degree from Universiti Tunku Abdul Rahman in 2011. In 2008 and 2015, respectively, he underwent research attachment in the University of Oxford (UK) and Nippon Institute of Technology (Japan). He is the external examiner of Wawasan Open University. He is also the editor in chief of the i-manager's Journal on Digital Signal Processing. He has also been a guest editor for the Journal of Applied Environmental and Biological Sciences and Journal of Fundamental and Applied Sciences. When working in Intel Corporation, he was involved in the design of the Pentium IV PSC and Celeron NWD-V microprocessors. This earned him four Kudos awards from Intel Microelectronics. He has also been given the university teaching excellence award and 16 research grants. He has published close to 100 scientific articles, which include refereed journal and conference papers, books and book chapters.

Humaira Nisar has a BS degree in Electrical Engineering from the University of Engineering and Technology, Lahore, Pakistan; MS degree in Nuclear Engineering from Quaid-i-Azam University, Islamabad, Pakistan; and another MS degree in Mechatronics and PhD degree in Information and Mechatronics from Gwangju Institute of Science and Technology, Gwangju, South Korea. She has more than 15 years of research experience. Currently, she is working as an Associate Professor in the Department of Electronic Engineering, Universiti Tunku Abdul Rahman, Malaysia. She is also the head of Programme for the Master of Engineering Science Programme. She is a senior member of IEEE. Her research interests include signal and image processing, biomedical imaging, brain signal and image analysis and image analysis for wastewater treatment. She has published a number of international journal and conference papers. She has also served on technical committees of various conferences and journals.

Contents

Preface

In this book, a variety of topics related to Very-Large-Scale Integration (VLSI) is extensively discussed. The topics encompass the physics of VLSI transistors, the process of integrated circuit (IC) design and fabrication and the applications of VLSI devices. The book is intended to provide information on the latest advancement of VLSI technology to researchers, physicists as well as engineers working in the field of semiconductor manufacturing and VLSI design.

Chapter 1 gives an overview of VLSI-integrated circuit devices. A brief historical development of the transistor and integrated circuits is first presented. This is then followed by an introduction of the field effect transistors and of the technology progression driven by Moore's law, which necessitates the evolution of MOSFETs to FinFETs. A concise walkthrough of the VLSI manufacturing process—from design to fabrication and finally to packaging—is also illustrated at the end of the chapter.

In Chapter 2, the historical evolution of the hot carrier degradation mechanisms and their physical models are reviewed. An energy-driven hot carrier ageing model that can reproduce the 62-nm-gate-long hot carrier degradation of the transistor is verified. A long-term hot carrier-resistant circuit design can be realized via optimal driver strength controls. The central role of the V_{GS} ratio is emphasized during practical case studies on CMOS inverter chains and a DRAM word line circuit. Negative bias temperature instability (NBTI) mechanisms are also reviewed and implemented in a hydrogen reaction-diffusion (R-D) framework. The R-D simulation reproduces time-dependent NBTI degradations interpreted into interface trap generation, , with a proper power-law dependency on time. The experimental evidence of pre-existing hydrogen-induced Si-H bond breakage is also proven by the quantifying R-D simulation. The analysis shows that a low-pressure EOL anneal can reduce the saturation level of NBTI degradation, which is believed to be caused by the outward diffusion of hydrogen from the gate regions. It therefore prevents further breakage of Si-H bonds in the silicon-oxide interfaces.

As VLSI technology advances, power consumption in a chip becomes an essential factor, which must be accounted for during IC design. Hence, power optimization is necessary at all levels of the design process. Chapter 3 looks into various design methodologies that could be applied to achieve the concept of low-power designs.

Chapter 4 discusses in detail highly pure refractory metals, which are used in VLSI thin-film metallization. Cast targets of highly pure refractory metals, such as W, Mo, Ti, Ta, Co, etc., and their compounds can be produced by means of a set of vacuum metallurgical techniques, i.e., vacuum high-frequency levitation, EB floating zone melting, EB melting and electric arc vacuum melting, as well as chemical purifying by ion exchange and halides. The cast refractory metal targets are extremely pure and chemically homogeneous. For magnetron sputtering and laser ablation, the cast silicide targets are produced. The study reveals the possibilities and conditions of depositing the silicides and titanium-tungsten barrier layers by both laser evaporation and magnetron sputtering. The physical and structural parameters as well as a trace impurity composition of sputtered metals and deposited thin films are studied by grazing-beam incidence X-ray diffraction, Auger electron spectroscopy, Rutherford backscattering of helium ions, mass spectrometry with inductively coupled plasma, etc.

In Chapter 5, the design of the operational amplifier (op-amp) and its application at the front-end circuitry of a mixed-signal IC are illustrated. The chapter focuses on the design of the op-amp based on different compensation schemes and by incorporating negative Miller compensation. The op-amps are designed to operate at low voltage levels. The g_m/I_D ratio design approach is used to determine the transistor sizes. The op-amp performance is simulated using Cadence Spectre. Open-loop frequency responses are given emphasis during validation.

Chapter 6 presents the design of second-order analogue notch filters based on floating active inductors and conventional second-order notch LCR prototypes. The designs are validated using Cadence Spectre, 0.18 micron library. High-order notch filters are also reviewed. The simulation results of a sixth-order notch filter is shown to have ideal polynomial and ideal pole and zero position.

Kim Ho Yeap
Associate Professor
Department of Electronic Engineering
Faculty of Engineering and Green Technology
Universiti Tunku Abdul Rahman, Malaysia

Humaira Nisar
Associate Professor
Department of Electronic Engineering
Universiti Tunku Abdul Rahman, Malaysia

Section 1

Introduction

Chapter 1

Introductory Chapter: VLSI

Kim Ho Yeap and Humaira Nisar

Additional information is available at the end of the chapter

http://dx.doi.org/10.5772/intechopen.69188

1. Introduction

Back in the old days about 40 years ago, the number of transistors found in a chip was, even at its highest count, less than 10,000. Take, for example, the once popular Motorola 6800 microprocessor developed in the mid 1970s. Fabricated based on the 6.0-μm feature size, the 6800 consisted of merely 4100 transistors in it. Nowadays, the number of transistors in a very large-scale integration (VLSI) [or some refer to it as the super large-scale integration (SLSI)] chip may possibly reach 10 billion, with a feature size smaller than 15 nm.

There is little doubt that the electronics world has experienced a significant advancement for the past 50 years or so and this, to a large extent, is due to the rapid technology improvement in the performance, power, area, cost and 'time to market' of an integrated circuit (IC) chip. To provide readers with an overall view of VLSI, this chapter gives a concise but complete illustration on the historical evolution, design and development of VLSI-integrated circuit devices.

2. A brief history

When transistors were first introduced in early 1900s, they were actually made of vacuum tubes. These transistors were relatively large in size and cumbersome to be used. In December 1947, however, three physicists working in the AT&T Bell laboratory attained a remarkable breakthrough in their design, changing the perception one used to have on transistors. Dr. John Bardeen, Dr. Walter Houser Brattain and Dr. William Bradford Shockley, Jr., invented the first point-contact semiconductor transistor using germanium. As shown in **Figure 1**, the point-contact transistor comprised an n-type germanium block and two gold contacts (i.e. the emitter and collector leads) placed in close proximity. When a small current was applied to one of the contacts, the output current at the other contacts was amplified.

Being much smaller in size, consuming much lower power, operating at relatively lower temperature and giving quicker response time, the semiconductor transistor is clearly more

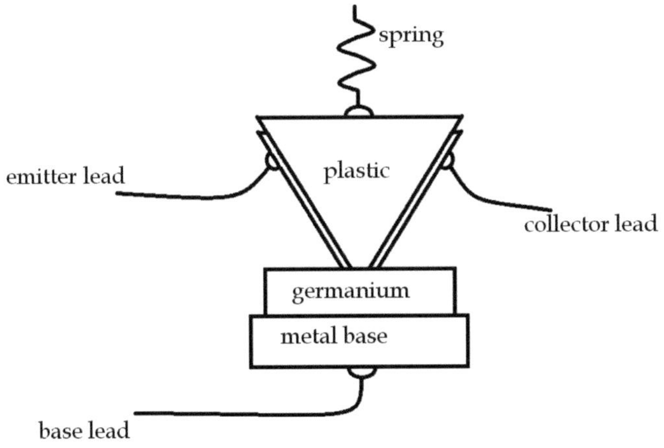

Figure 1. An early model of the point-contact transistor.

superior to its conventional vacuum tubes brethren. It was these advantages and its viability that resulted in the replacement of vacuum tubes by the solid-state electronic devices. The rapid widespread usage of the semiconductor transistors in electronic circuits has triggered a dramatic revolution in the electronic industries, kicking off the era of semiconductor. Because of this significant contribution, Bardeen, Brattain and Shockley shared the Nobel Prize in Physics in 1956.

It may be worth noting that when this germanium solid-state device was initially introduced, it was not coined the term 'transistor'. Instead, it was generally referred to as the 'semiconductor triode'. According to the 'Memorandum For File' of Bell Telephone Laboratories [1], six names had been proposed for the device, namely 'semiconductor triode', 'surface states triode', 'crystal triode', 'solid triode', 'iotatron' and 'transistor'. Although the device had initially been referred to as the 'semiconductor triode', the word 'transistor' (which originates from the abbreviated combinations of the words 'transconductance' and 'varistor') had ultimately turned out to be the winner of the internal poll [1].

The first commercially available silicon transistors were manufactured by Dr. Gordon Kidd Teal in 1954. Since silicon gives much better performance than germanium transistors, the substrate material for transistors was gradually changed to silicon. In 1955, the first diffused silicon transistor made its appearance. To reduce the resistivity of the collector, the transistor with an epitaxial layer added onto it was developed in 1960. It was also in the same year the planar transistor was proposed by Dr. Jean Amedee Hoerni [2].

In 1958, Jack St. Clair Kilby who was then an engineer in Texas Instruments successfully developed the first integrated circuit. The device was just a simple 0.5-inch germanium bar, with a transistor, a capacitor and three resistors connected together using fine platinum wires. About a year later in 1959, Dr. Robert Norton Noyce from Fairchild Camera (also one of the co-founders of Intel Corporation) invented independently his own integrated circuit chip. The interconnection in Noyce's 4-inch silicon wafer was realized by means of etching the aluminium film which was first deposited onto a layer of oxide [2]. Both Kilby and Noyce shared the patent right for the invention of the integrated circuit. In 2000, Kilby was awarded the Nobel Prize in Physics 'for his part in the invention of the integrated circuit'.

Since the advent of the semiconductor transistor and the demonstration on the workability of the integrated circuit chip about some 70 years ago, the electronic industries have been prospering hitherto. Electronic devices are now closely interwoven with human's life. They have, in many aspects, become indispensable to mankind. Indeed, one can easily find traces of electronic circuitries integrated into areas which intertwine seamlessly with the fabric of mankind's living hood. Some of these areas include transportation, telecommunication, security, medicine and entertainment, just to name a few.

3. Moore's law

In April 1965, one of the co-founders of Intel Corporation, Dr. Gordon Earle Moore, predicted that the number of components (i.e. any electronic components which include not just transistors but capacitors, resistors, inductors, diodes, etc. as well) in an integrated circuit would double every year [3]. Ten years later in 1975, he revised his prediction to a doubling of every 2 years. Moore's prediction, which is more commonly known as Moore's law nowadays, has been widely used in the semiconductor and microelectronic industries as a tool to predict the increase of components in a chip for the coming generations [4]. To date, Moore's law has been proven to have held valid for more than half a century. **Table 1** depicts the progressive trend of the integration level for the semiconductor industry. It can be observed from the table that the number of transistors that can be fabricated in a chip has been growing continuously over the years. In fact, this growth has complied closely with Moore's law. To distinguish the increase of transistors in every 10 years, each era is designated a name, that is, the SSI, MSI, LSI, VLSI, ULSI and SLSI eras. During the VLSI era, a microprocessor was fabricated for the first time into a single integrated circuit chip. Although this era has now long passed, the VLSI term is still being widely used today. This is partly due to the absence of an obvious qualitative leap between VLSI and its subsequent ULSI and SLSI eras, and partly, it is also because IC engineers and experts working in this field have been so used to this term that they decided to continue adopting it.

Integration level	Year	Number of transistors in a chip
Small-scale integration (SSI)	1950	Less than 100
Medium-scale integration (MSI)	1960	Between 100 and 1000
Large-scale integration (LSI)	1970	Between 1000 and 10,000
Very large-scale integration (VLSI)	1980	Between 10,000 and 100,000
Ultra large-scale Integration (ULSI)	1990	Between 100,000 and 10,000,000
Super large-scale integration (SLSI)	2000	More than 10,000,000

Table 1. Integration level of an integrated circuit chip.

4. The field effect transistors

Today, the transistors fabricated in an IC device are mostly metal oxide semiconductor field effect transistors (MOSFETs). The earliest paper describing the operation principle of a MOSFET can be traced back to that reported in Julius Edgar Lilienfeld's patent in 1933 [5]. In 1959, Dr. Dawon Kahng and Dr. Martin M. (John) Atalla at the Bell Telephone Laboratories successfully invented the MOSFET [6]. In 1963, two engineers from the Radio Corporation of America (RCA) Princeton laboratory, Dr. Steven R. Hofstein and Dr. Frederic P. Heiman, presented the theoretical description on the fundamental nature of the silicon planar MOSFET [7]. In the same year, Dr. Frank Marion Wanlass of Fairchild Semiconductor invented the first complementary metal oxide semiconductor (CMOS) logic circuit [8].

4.1. The MOSFET

The MOSFET is basically a device that operates like a switch or an amplifier in electronic circuits. **Figure 2** depicts the basic structure of the MOSFET. The device consists of four terminals, namely the drain (*D*), source (*S*), gate (*G*) and substrate or bulk (*B*) terminals. Basically, the device is composed of three layers—a poly-silicon layer (i.e. the gate terminal), an oxide layer (i.e. the gate oxide) and a single-crystal semiconductor layer (i.e. the substrate). In the early days, the gate terminal was made of aluminium. It is from these three layers of materials that the FET device derived its name. In mid 1970s, however, the gate material was replaced with polysilicon. The high-temperature stability of the polysilicon gate is used as a mask to form the self-aligned source and drain terminals via ion implantation, rendering higher accuracy for the formation of these two terminals. Although the gate today is no longer made of aluminium, the term MOSFET has been so widely accepted that it stays until today.

The operation principle of a MOSFET is actually quite simple. When a voltage is applied in between the drain and source terminals, a conducting channel is required to be formed between the two terminals to close the circuit (i.e. to allow current to flow). A voltage connected to the gate terminal acts like a switch. Given sufficient magnitude (and the correct

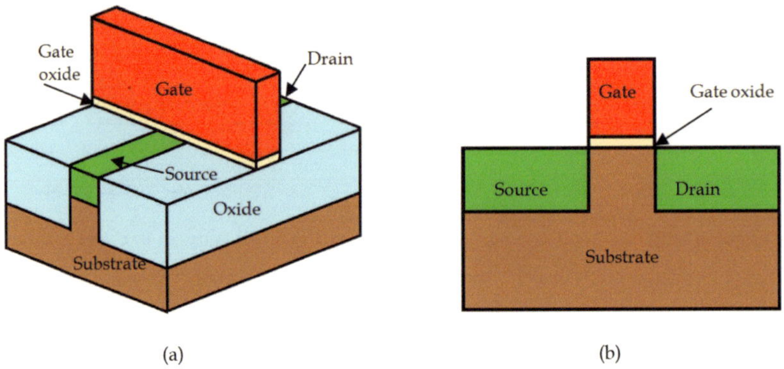

Figure 2. The (a) basic structure and (b) cross section of a MOSFET.

polarity), the gate voltage is able to attract carriers to the gate oxide-substrate interface, forming a channel which connects the source and drain terminals.

A MOSFET can be categorized into two types, depending on the dopants of the drain and source terminals, as well as the substrate. When both the drain and source terminals, in a p-type substrate, are heavily doped with donator ions (such as phosphorous or arsenic), a negative channel is to be formed in between them to conduct current. On the other hand, when both terminals, in an n-type substrate, are heavily doped with acceptor ions (such as boron), a positive channel is to be formed. The former device is therefore known as a negative-channel MOSFET or an NMOS transistor, while the latter is known as a positive-channel MOSFET or a PMOS transistor. **Figure 3** shows the circuit symbols of both PMOS and NMOS transistors.

The size of a MOSFET transistor is measured by the gate length, which is also commonly known as the feature size or feature length *L*. The feature size *L* has been shrinking tremendously over the years. Transistors with the size of 50 μm in the 1960s have been scaled down to less than 15 nm in 2017. The reduction of size allows a higher density of transistors to be fabricated in a single die. Overseen by the Taiwan Semiconductor Industry Association (TSIA), the United States Semiconductor Association (SIA), the European Semiconductor Industry Association (ESIA), the Japan Electronics and Information Technology Industries Association (JEITA) and the Korean Semiconductor Industry Association (KSIA), the International Technology Roadmap of Semiconductor (ITRS) is produced to forecast how the technology is expected to evolve. The purpose of the ITRS is to ensure healthy growth of the IC industries. **Table 2** lists the progressive reduction of the feature size published in ITRS 2.0 [9].

4.2. The FinFET

As the feature size reduces to the submicron regimes, fields at the source and drain regions may become comparatively high, and this may give certain adverse effects to the charge

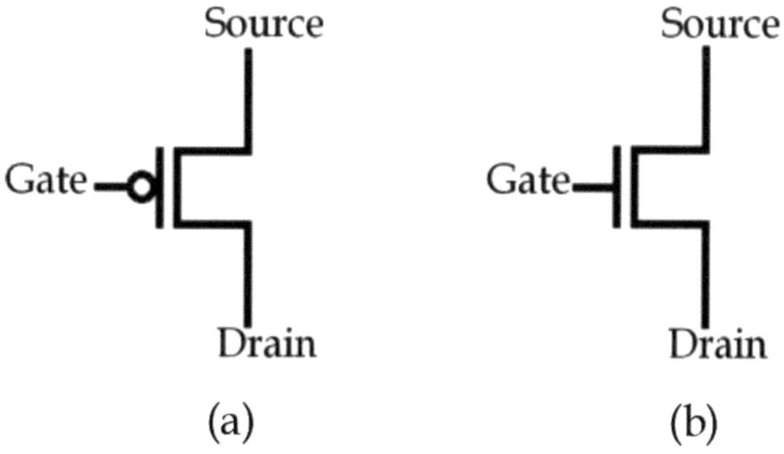

Figure 3. The symbol of (a) a PMOS transistor and (b) an NMOS transistor.

Physical gate length	Year						
	2015	2017	2019	2021	2024	2027	2030
High-performance logic (nm)	24	18	14	10	10	10	10
Low-performance logic (nm)	24	20	16	12	12	12	12

Table 2. Forecast of gate length by ITRS.

distribution. Some of the examples of these short-channel effects are the threshold voltage roll-off in the linear region, drain-induced barrier lowering (DIBL) and bulk punch-through [10]. To suppress these effects, additional steps, such as the introduction of retrograde well, lightly doped drain, halo implantation, and so on, have been introduced to the IC fabrication process [11]. As the device continues to shrink, however, curbing the short-channel effects turns out to be a strenuous task. When the feature size approaches the sub-nanometre range (i.e. 90 nm and below), static leakage current due to the short-channel effects has become a serious problem.

When the technology node reached 22 nm in 2011, Intel Corporation announced the fabrication of the tri-gate transistor, replacing the conventional planar MOSFET. More commonly known as the FinFET, this device has a three-dimensional transistor structure, as shown in **Figure 4**. From the figure, it is clear that a FinFET is named so because of the protruding source/drain terminals from its substrate surface, which closely resemble the fins of a fish. Since the gate wraps around the inversion layer, FinFETs provide higher current flow from source to drain. This feature also allows better control of the current flow—it reduces current leakage considerably when the device is at its 'off-state' and minimizes short-channel effects at its 'on-state'. Since the device has lower threshold voltage than the planar MOSFET, a FinFET can also operate at a lower voltage. In other words, the new device shows less leakage, faster switching and lower power consumption. However, certainly, the efficiency improvement

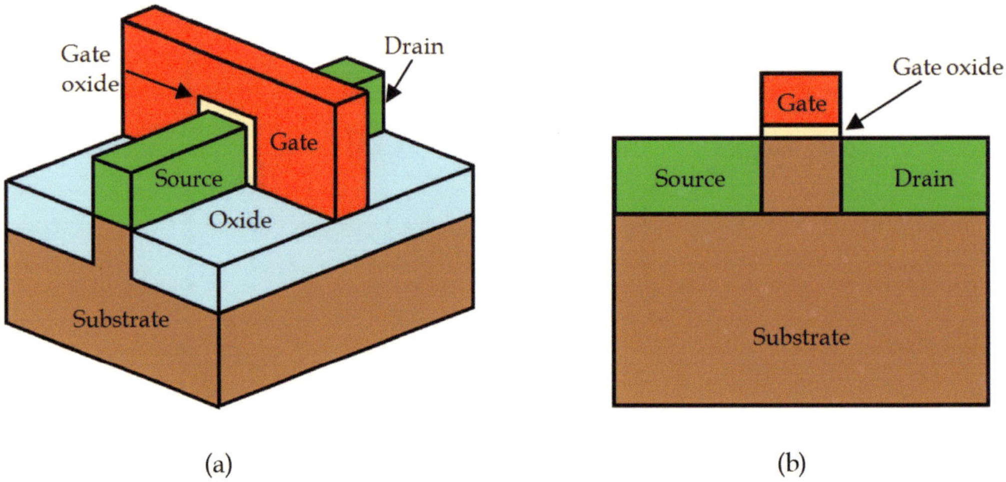

Figure 4. The (a) basic structure and (b) cross section of a FinFET.

found in the FinFET comes at the expense of increased fabrication complexity. The introduction of additional fabrication steps is inevitable in order to form the fin-like structure.

5. VLSI design flow

Generally, the design process of a VLSI chip involves three stages namely the (i) behavioural, (ii) logic circuit and (iii) layout representations. At each of this stage, verification is to be performed at the end before proceeding to the next. Hence, it is common to have repetitions and iterations in the processes [12].

5.1. Behavioural representation

Behavioural representation is the first step of the entire VLSI design flow. At this stage, it is important to specify the functionalities of the device and how it is going to communicate with the exterior. The design architecture is to be drawn panned out. A hardware description language (HDL) such as Verilog HDL or VHDL is used to define the behaviour of the device.

5.2. Logic circuit representation

After the HDL codes are successfully simulated, functional blocks from standard cell libraries are used to synthesize the behavioural representation of the design into logic circuit representation. Once the design is verified, the gate level netlist is generated. The netlist is necessary in order to develop the layout of the design.

5.3. Layout representation

At the final stage, the physical layout of the design is created. The process starts with floor planning which defines the core and routing areas of the chip. In order to optimize the design, the building blocks are arranged and orientated at their best locations. This process is known as placement. Once this is completed, a routing process is performed to interconnect the building blocks.

6. IC fabrication

To fabricate the chip, the layout is sent to a fab or a foundry. In a fab, a single-crystal semiconductor ingot is first grown. Wafers are then sliced from the ingot. The layout is printed onto the dice in each wafer. In the initial step of chip fabrication, the active regions or wells for the NMOS and PMOS transistors are first formed at the substrate. In order to separate the transistors, an oxide layer is subsequently deposited in between each neighbouring well. Transistors are then built at each active region. The primary processes to form a transistor include the growth of the gate oxide layer, the deposition of the poly-gate, and the doping of the source and drain regions. In the final fabrication step, the transistors are interconnected in accordance to the layout of the design. In a nutshell, the process of chip fabrication can be broadly separated into four stages: (i) well formation, (ii) device isolation, (iii) transistor making and (iv) interconnection [13]. Although the walkthrough may appear straight forward, it is, in practical, complicated and laborious. To fabricate a VLSI chip, the die has to undergo repetitive thermal processes (such as oxidation, diffusion, annealing, etc.), lithography, ion implantation, etch, dielectric film deposition (such as chemical vapour deposition or CVD), chemical mechanical polishing (CMP) and metallization [13].

7. IC packaging

To protect the chip from harsh external environment (e.g. being exposed to UV light or moisture or being scratched), it is essential to encapsulate the chip in a package. The three most commonly used techniques for packaging are (i) wirebonding, (ii) flip-chip and (iii) tape-automated bonding (TAB) [4]. Once the chip is carefully packaged, it is then ready to be released to the market.

Author details

Kim Ho Yeap* and Humaira Nisar

*Address all correspondence to: yeapkimho@gmail.com

Tunku Abdul Rahman University, Jalan Universiti, Bandar Barat, Kampar, Perak, Malaysia

References

[1] Arsov GL. Celebrating 65th anniversary of the transistor. Electronics. 2013;**17**:63-70. DOI: 10.7251/ELS1317063A

[2] Lukasiak L, Jakubowski A. History of semiconductors. Journal of Telecommunications and Information Technology. 2010;**1**:3-9. DOI: 10.1088/0031-9120/40/5/002

[3] Moore GE. Cramming more components onto integrated circuits. Electronics. 1965;**38**:14-117. DOI: 10.1109/N-SSC.2006.4785860

[4] Yeap KH. Fundamentals of Digital Integrated Circuit Design. 1st ed. Central Milton Keynes: Authorhouse; 2011

[5] Lilienfeld JA. Method and apparatus for controlling electric currents. U. S. Patent No. 1745175A (Filed October 8, 1926, issued January 28, 1930)

[6] Kahng D. Electric field controlled semiconductor device. U. S. Patent No. 3, 102, 230 (Filed 31 May 31, 1960, issued August 27, 1963)

[7] Hofstein S R, Heiman, FP. Silicon insulated-gate field-effect-transistor. Proceedings of IEEE. 1963;**51**:1190-1202

[8] Wanlass SM, Sah CT. Nanowatt logic using field-effect metal-oxide semiconductor triodes. In: Proceedings of the IEEE Conference on Solid-state Circuits Conference. Digest of Technical Papers; 1963; Philadelphia, PA, US. p. 32-33

[9] International Technology Roadmap for Semiconductors 2.0 [Internet]. 2015. Available from: https://www.semiconductors.org [Accessed: 3 April 2017]

[10] Sze SM. Semiconductor Devices: Physics and Technology. 2nd ed. New York, NY: John Wiley and Sons; 2002

[11] Ahmad I, Ho YK, Majlis BY. Fabrication and characterization of a 0.14 μm CMOS device using ATHENA and ATLAS simulators. International Scientific Journal of Semiconductor, Physics, Quantum Electronics, and Optoelectronics. 2006;**9**: 40-44. DOI: https://doi.org/10.15407/spqeo

[12] Kang S–M, Leblebici Y. CMOS Digital Integrated Circuits: Analysis and Design. Singapore: McGraw-Hill; 2005

[13] Xiao H. Introduction to Semiconductor Manufacturing Technology. Upper Saddle River, New Jersey: Prentice Hall; 2001

Section 2

Physics of MOSFET's

Chapter 2

Transistor Degradations in Very Large-Scale-Integrated CMOS Technologies

Chang Yeol Lee

Additional information is available at the end of the chapter

http://dx.doi.org/10.5772/intechopen.68825

Abstract

The historical evolution of hot carrier degradation mechanisms and their physical models are reviewed and an energy-driven hot carrier aging model is verified that can reproduce 62-nm-gate-long hot carrier degradation of transistors through consistent aging-parameter extractions for circuit simulation. A long-term hot carrier-resistant circuit design can be realized via optimal driver strength controls. The central role of the V_{GS} ratio is emphasized during practical case studies on CMOS inverter chains and a dynamic random access memory (DRAM) word-line circuit. Negative bias temperature instability (NBTI) mechanisms are also reviewed and implemented in a hydrogen reaction-diffusion (R-D) framework. The R-D simulation reproduces time-dependent NBTI degradations interpreted into interface trap generation, ΔN_{it} with a proper power-law dependency on time. The experimental evidence of pre-existing hydrogen-induced Si–H bond breakage is also proven by the quantifying R-D simulation. From this analysis, a low-pressure end-of-line (EOL) anneal can reduce the saturation level of NBTI degradation, which is believed to be caused by the outward diffusion of hydrogen from the gate regions and therefore prevents further breakage of Si–H bonds in the silicon-oxide interfaces.

Keywords: hot carrier injection (HCI), hot carrier degradation (HCD), hot carrier-resistant design, negative bias temperature instability (NBTI), reaction-diffusion (R-D) of hydrogen

1. Introduction

Since the concept of an integrated circuit was first proposed by Jack Kilby in 1958, and a first version of a self-aligned poly-silicon gate CMOS-integrated circuit was fabricated at Fairchild® in 1968, integrated circuit technology has led to unprecedented thriving and prosperity in the electronic industry for the last half century. The initial integration of a number of transistors started with only a few tens in a circuit, which we call small-scale integration (SSI) has today expanded to a few billions, called very large-scale integration (VLSI) or ultra large-scale integration (ULSI). The annual growth rate of the number of transistors per IC has followed a well-known formula, Moore's Law, which indicates that the density of devices per chip doubles every 18 months, that is, the population of transistors in a chip increases by 1000 times every 15 years. Although in the beginning it was a merely an observation, it evolved into a de facto mandatory target for cutting-edge technology developers. For example, the Intel® CPU transistor count has faithfully followed Moore's Law for four decades (1971–2012) (**Figure 1**). Specifically, their newest microprocessor, Ivy Bridge Core i7™, possesses 1,400,000,000 transistors [1]. It is a 609,000× increase in transistor count from its 1971 version. Such an exponential increase in integration number is attributed to a series of successes in shrinking feature size. The benefit of scaling is obvious: more integrated transistors enable more sophisticated data-driven operations and less switching delays per logic gate, thereby enhancing data transaction bandwidth. More data with enhanced speed play a decisive role in the rapid growth of the information and communication industry.

During the continuous pursuit of scaling, the following inherent issues have arisen:

- The challenge to sustain photolithographic pattern fidelity and critical dimension (CD) uniformity becomes profound as dimension scaling and integration levels increase.
- As transistor gate length is shrunk, electric field strength inside of a transistor increases and more degradation may occur in devices. Vertical and lateral e-fields can be mitigated by reduction of bias voltage, V_{DD}. Its minimum level tends to be limited by the minimum threshold voltage of logic gates defined by distinguishable high states against thermal noise and permissible off-leakage currents. Since the V_{DD} limit of around 1.0 V is already achieved in state-of-the-art technologies, inevitable increases of internal e-fields inside of devices may define the practical limit of technology scaling.

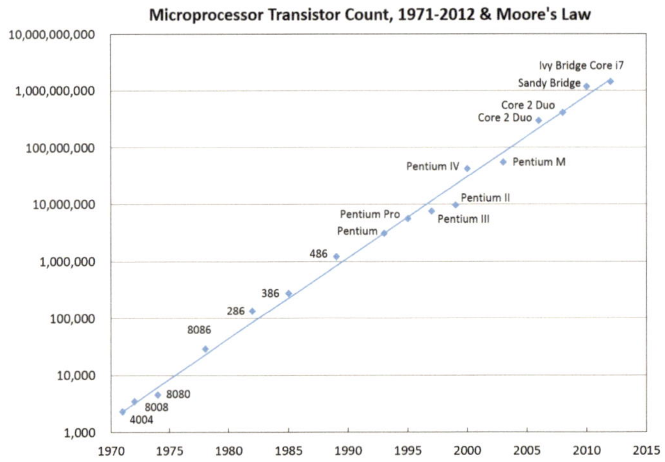

Figure 1. Intel® microprocessor transistor count evolution between 1971 and 2012. For the last 40 years, it has consistently behaved as predicted by Moore's Law. (Source: Intel Corp, Web Page [1]. Figure 1 is a graphical translated version, which was originally expressed in a tabular form in [1].)

The first issue is directly related to how short are the wavelengths of photolithographic light sources we employ. A 193-nm ArF light source with immersion ambience is known to have 40-nm patterning capability as its best performance. Since a feasible solution for a new light source with a lesser wavelength than ArF has not yet been found, a complicated combination of photolithography and etching processes, for example, a double-spacer pattern technology (D-SPT), is employed to enable the latest 10-nm range patterns. Such a complicated combination of critical steps may cause a sizable variation of CDs. As a consequence, much expertise and extensive trials and corrections are required to achieve pattern optimization, which becomes the confidential property of cutting-edge companies.

So solving the first issue relies in large part on skill and trial-and-error correction processes. By contrast, the second issue is related largely to scientific analysis. When bias is applied to a scaled MOSFET, a localized e-field is established in the drain side, which can accelerate mobile carriers (electrons or holes) passing this region. Some of the accelerated carriers can trigger an avalanche multiplication process, which increases the possibility of generating energetic carriers that can surmount the energy barrier between silicon/silicon dioxide or cause damage in Si/SiO_2 interface. Energetic carriers or "hot carriers" can also be generated by energy exchange during carrier-carrier-scattering processes. This kind of device degradation mechanism is called "hot carrier injection (HCI)," or "hot carrier degradation (HCD)" and is regarded as a typical degradation mechanism driven by high lateral e-fields or V_{DD}. The most efficient prevention of this kind of degradation is reducing V_{DD}. Large efforts have been devoted in device and circuit research to develop power-efficient and degradation-aware low V_{DD} transistors and circuit solutions. Despite this effort, there still inevitably remain high voltage needs in some specific applications, like word-line decoders in dynamic random access memory (DRAM) circuits.

Dynamic Random Access Memory (DRAM) is one of the most popular memory devices featuring high data read/write speed with low bit cost. Compact placement of a single-bit storage capacitor and its switch transistor composes a DRAM cell. To avoid large off-leakage caused by high e-fields and thereby insufficient data retention capability, a three-dimensional (3-D) recess-channel scheme has recently been developed to reduce the e-field strength by extending the channel length. Although sufficient data retention time can be achieved by the 3-D recess-channel structures, reduction of the channel conductance of the long channel length expensively undermines the access speed. Non-scaled gate voltage can compensate for this loss. Such a decoupling from scaling rules (planar dimensions scale down, whereas gate voltages do not) may cause HCD issues in cell gate bias-pumping voltage (V_{PP}) circuits. Since HCD is ascribed to the high electric field and/or high gate voltage, the mitigation strategy largely relies not only on device internal structure and doping profiles but also on the circuit and layout strategy. More detailed descriptions based on practical case studies and some general guidelines for the HCI-resistant circuit design can be found in the next section.

Lateral (channel length) shrinkage should be indispensably coupled with vertical (gate oxide thickness) shrinkage to maintain "long channel-like" transistor characteristics. The key enabler of vertical scaling is the superb electrical and material properties of silicon dioxide. As to silicon dioxide, only about 10 stacks of molecules can provide good isolation under 5.5–6.0 MV/cm of electric field intensity or can sustain "off-characteristics" of the few tens of nanometer-scaled MOSFET. Despite its stability, modern plasma-intensive fabrication processes can induce multiple charging in the gate electrodes, which can generate a number of silicon-oxide bond breakages. Most bond breakages can be passivated by end-of-line (EOL) hydrogen or deuterium passivation steps in order to electrically deactivate the dangling bonds. In this circumstance, another kind of device degradation mechanism can be triggered: negative bias temperature instability (NBTI) can be activated by moderated gate bias and temperatures applied in p-channels where abundant inverted holes and hydrogen-passivated silicon-dangling bonds exist. Although its mechanism is still not completely understood, atomic and/or molecular hydrogen reactions and diffusions associated with passivated Si/SiO_2 interfaces are widely accepted to define how and how much degradation takes place. In Section 3, a quantitative analysis is provided based on the reaction-diffusion of hydrogen simulation. A mitigation strategy for the long-term NBTI degradation is also suggested during the analysis.

In this chapter, studies on the most typical scaled-down-related device reliability issues, HCD in NMOS and NBTI in PMOS, are presented with practical case studies to attempt to broaden the reader's knowledge of device degradation and its impact on advanced CMOS scaling.

2. Hot carrier degradation

2.1. Historical review

Hot carrier degradation (HCD) is one of the typical wear-out degradation mechanisms that causes catastrophic failures in systems. This kind of failure may implosively trigger irreversible and unrecoverable damage in systems. Readers can find typical cases of HCD failure syndromes and their impacts on complete products in Ref [2]. Many investigations have been conducted to reveal the transistor degrading hot carrier generation mechanism. The first successful theory structure was announced as the lucky electron model (LEM) suggested by Hu et al. in 1985 [3]. The LEM is regarded as a classic theory and has been widely used so far because it allows the depiction of a clear image for hot carrier generation and its role in creating interface traps. It focuses on e-field-driven hot carrier generation. A quasi two-dimensional analysis of Poisson's equation derives the exponential shape of e-fields in the velocity saturation region (VSR) and a sharp peak of e-fields built in front of the neutral drain region [4]. All the energy gain processes are assumed to be concentrated in the peak e-field spot where lucky electrons are generated by re-directional impact ionizations with Si-lattices. As a result, lucky electrons can surmount the Si/SiO_2 energy barrier (3.2 eV) and generate the Si interfacial traps, N_{it} [3]. On this basis, the following interface generation rate, r_{it}, was derived:

$$\begin{aligned} r_{\text{it,LEM}} &= C_1 \cdot I_D \cdot \left(\frac{I_B}{I_D}\right)^{\Phi_{\text{it}}/\Phi_{\text{ii}}} \\ &= C_1 \cdot I_D \cdot \left[A \cdot (V_D - V_{\text{DSAT}}) \cdot \exp\left(\frac{-B\ell}{V_D - V_{\text{DSAT}}}\right)\right]^{\Phi_{\text{it}}/\Phi_{\text{ii}}} \end{aligned} \tag{1}$$

where I_B and I_D are substrate current and drain current, respectively, Φ_{ii} is the impact ionization threshold energy (1.3 eV for electrons) and Φ_{it} is the interface state generation threshold energy (3.7 eV for electrons), ℓ is the characteristic length of VSR, and A, B, and C are constants, respectively. The power-law exponent, Φ_{it}/Φ_{ii}, is calculated to be 2.8 and this approximately matches experimental results that guarantee that the formula captures the correct image of HCD.

Transistor degradations mean that threshold voltages shift, mobility decreases, and drain-extrinsic resistances increase, all of which are ascribed to the interface trap generation. An isotope effect found by the scanning tunneling microscope (STM) method [5] and HCD experiments using hydrogen and deuterium-annealed samples [6] reveals that the dissociation of hydrogen from the interfacial Si–H bonds by injected energetic electrons can lead to unrecoverable degradations. The interface trap generation rate is empirically expressed as

$$\Delta N_{it}(t) \propto (r_{it} \cdot t)^{n} \tag{2}$$

where n is a time exponent that has been known to be around 0.5, which can be derived from a hydrogen diffusion-limited process [3]. An assumption of high diffusivities of hydrogen in silicon dioxide and in polysilicon gate regions is required to describe the 0.5 dependency. More specifically quickly removing the hydrogen from the interface and therefore also the repassivation process cannot dominantly influence the whole hydrogen reaction-diffusion process. Contradictable findings have been also reported in PMOS-negative bias temperature instability (NBTI) research [7]. Fast-diffused hydrogen in the SiO_2 region slows down in the silicon-nitride interface and in the polysilicon region due to small diffusion constants in those regions. This results in an accumulation of hydrogen in the SiO_2 region, which strikes a balance between dissociation and repassivation of silicon-dangling bonds. As a consequence, the interface state generation rate decreases to produce a smaller n (1/4–1/6). The discrepancy of time exponents between NMOS HCI and PMOS NBTI can be ascribed to the difference in the stressed area (only localized to the peak e-field spot in HCI vs. the whole gate oxide area in NBTI) and its influence on hydrogen diffusion profiles: an increasingly wider diffusion front of hydrogen in NMOS HCD enhances the ΔN_{it} rate more than that of PMOS NBTI where consistent one-dimensional diffusion of hydrogen occurs [8]. Furthermore, asymmetric behaviors between NMOS and PMOS (a large amount of degradation is quickly recovered when the stress biases are removed in PMOS, while no substantial recovery takes place in NMOS) imply the different nature of the Si-H dissociation produced by cold holes injected during NBTI stress and by hot electrons injected during HCI stress. The existence of deep-level hole traps (DLHT) [9, 10] was proposed to draw a plausible picture of the asymmetric behavior. In the author's opinion, more studies are still needed to reveal the underlying physics for a comprehensive understanding.

As the e-field-driven (and consequently the applied voltage-driven) LEM reveals the HCD mechanism to be successful, it also instigates a voltage scale-down from 3.3–5.0 to 1.8–2.5 V in the shrunk gate length transistors in an effort to avoid HCD risk. However, HCD still remains against expectation in the 1.8–2.5-V regime. Neither hot electron injections into gate oxide (requires 3.2 eV at least) nor interface trap generations (requires 3.7 eV) may take place according to LEM because of insufficient driving voltage. A new hypothesis for the HCI generation mechanism, electron-electron scattering (EES), has been proposed to explain the hot carrier generation in medium V_{DD} conditions. This hypothesis has been accepted through numerous experimental verifications [11]. It involves an energy-exchanging electron-scattering process to generate hot electrons under moderate bias conditions. A doubling of its energy can be achieved when a perfect elastic collision between the excited electrons, 1.8–2.5 V, is sufficient to generate the interface-degrading hot electrons. The mathematical expression of EES implies these aspects as follows:

$$r_{\mathrm{it,EES}} = C_2 \cdot {I_D}^2 \cdot \left(\frac{I_B}{I_D}\right)^m. \tag{3}$$

Note that the power-law exponent of I_D changes from 1 as in Eq. (1) to 2 as in Eq. (3) reflecting a statistical interaction of two independent sources for EES. This secondly found hot carrier generation mechanism dominates in the sub-micrometer range-scaled MOSFETs whose drain currents have a range between 40 and 500 μA/μm with high V_{GS} drive [11, 12].

Further voltage scale-down to 1.0–1.2 V might extinguish any possibility of hot carrier generation via LEM or EES mechanisms. A newly developed Si–H bond breakage model has been proposed and demonstrated in deca-nanometer-scaled transistors [13]. Multivibrational hydrogen release (MVHR) is the third kind of mechanism, which is activated through high current injection (the minimum threshold is known to be 1.5 mA/μm), with weak voltage dependency. Since an electron can transfer its kinetic energy to the silicon lattice via optical phonon resonance, the multiple striking of electrons into Si–H bond can lead hydrogen to multiple jumps in its energy state to approach the bond-breaking threshold energy, E_B. This kind of hydrogen-dissociated process, namely multiple vibrational hydrogen release (MVHR), can be triggered by low-energy cold carriers in a sub-1-V-biased channel [13]. The quantum mechanical picture of the process is illustrated in **Figure 2**, and the mathematical expression was proposed to fit the experimental data as

$$r_{\mathrm{it,MVP}} = C_3 \cdot \left[{V_{\mathrm{DS}}}^{1/2}\left(\frac{I_D}{W}\right)\right]^{E_B/\hbar\omega} \cdot \exp\left(\frac{-E_{\mathrm{emi}}}{k_B T}\right) \tag{4}$$

where E_{emi} (~0.26 eV) is the barrier height of E_B from the highest energy state of bonded hydrogen. The unit resonance energy per single phonon excitation, $\hbar\omega$ (~0.075 eV), and threshold energy, E_B (~1.5 eV), defines the required number of the phonon excitation to be $\frac{E_B}{\hbar\omega} = 20$. Due to its tremendously strong power-law dependency on the drain current, a special caution should be paid not to let the transistor's drain current exceed the threshold of the third kind of HCD in any type of transistor operations including burn-in tests; otherwise, very quick wear-out failures may take place. The three kinds of hot carrier generation mechanisms are illustrated in **Figure 3** compared with a large set of experimental data.

To summarize the history of HCD mechanism finding, LEM dominates when $V_{\mathrm{DS}} \geq 3.0$ V. Energy-driven or current-driven multiple-particle (MP) mechanisms, EES for 40–500-μA/μm driving range and MVHR for even higher ranges, are subsequently developed. The maximum applied voltage and the minimum duty cycle of CMOS logic design guidelines have been

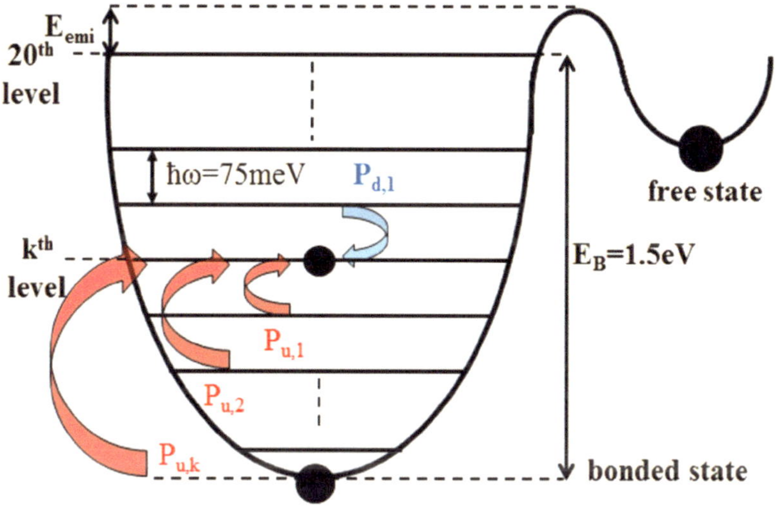

Figure 2. Schematic of Si-H potential well showing the processes increasing the occupancy of the kth level, coming from direct excitation, giving any number of energy quanta between 1 and k, and from the de-excitation of the $(k+1)^{th}$ level. $P_{u,i}$ and $P_{d,i}$ are, respectively, the probability of excitation and de-excitation, giving or losing i energy quanta. Reprinted with permission from Ref. [11], © 2013 IEEE.

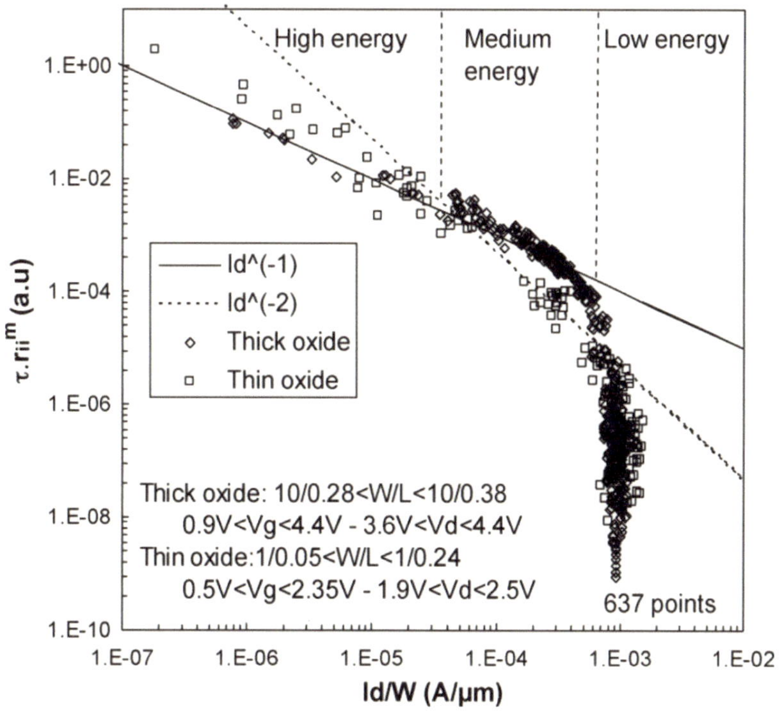

Figure 3. $\tau \cdot r_{ii}^{m}$ normalized lifetimes versus drain current showing three regimes of hot carriers. Reprinted with permission from Ref. [14], © 2007 IEEE.

made in strong awareness of the e-field-driven HCD. The maximum current-limiting constraints to prevent the current-driven HCD have not yet been made according to the author's knowledge. It might not be required since the ultimately scaled 3.8-nm-gate-long planar transistor demonstrates less than an (?) 1-mA/μm performance [15]. It is appreciably below the third limit. However, it can be exceeded by current boosting three-dimensional fin-gate structures.

Since the newest developed 3-D FinFETs have been announced reaching 1.0–1.5 mA/μm at V_{GS} of 0.75–0.8 V [16, 17], research work should concentrate on clarifying the risk of HCD in FinFET [18–20]. Inherent HCD risks in the FinFETs occur due to three reasons. First, the number of inversion electrons is increased by the surrounding three-dimensional gate overdriving ($V_{GS}-V_{TH}$), which supplies more electrons into the Si–H bond-breaking procedure through the multi-vibration mode. Second, the three-dimensional-surrounding gate introduces additional side interfaces between Si/SiO_2 by the fins, where additional Si–H bond breakage can occur. Furthermore, it can be enhanced when the gate is aligned to the (110) direction, the surface direction of the fin should also be (110) and the silicon surface density of the (110) plane is 1.4× larger than that of (100) plane [21]. Third, the three-dimensional surrounding of gate structures confines the heat dissipation only through the bottom-directional narrow body, thereby increasing the thermal resistance of the heat dissipation path. The lattice-carrier scattering generates heat that is referred to as "self-heating" and this increase of the lattice temperature is proportional to the thermal resistance. The temperature activation of HCD therefore becomes a critical reliability issue especially in high-current driving and poor heat dissipation devices like FinFETs [18]. A more detailed description of temperature dependency on hot carrier generations is found in the following section.

2.2. Temperature dependency on hot carrier generations

According to the LEM mechanism, carriers gain kinetic energy from the e-field, F, through free accelerated motion. The energy distribution of electrons is affected by the mean free path, λ, through,

$$f(E) = e^{-E/q\lambda F}. \tag{5}$$

Self-heating and/or ambient heating induces lattice vibrations that scatter the electrons to prevent gaining sufficient kinetic energy from triggering impact ionization. It can be assumed that λ decreases as the lattice temperature increases. As a result, higher energetic carriers can be generated at lower temperatures. **Figure 4** compares the long and high-biased (LH) transistors' and the short and low-biased (SL) transistors' HCI properties depending on temperature. That temperature dependency of the HCI lifetime and the substrate current follow the LEM picture in LH but not in SL suggesting that the LEM prediction is valid only in LH ranges but not in SL.

Monte Carlo simulation-based studies reveal that electron energy distribution function is composed of an e-field-driven main region and thermal tail [22, 23]. The knee voltage, $V_{EFF} = V_{DS} - V_{DSAT} + V_o$, separates two regions, where V_o is the voltage drop in the halo

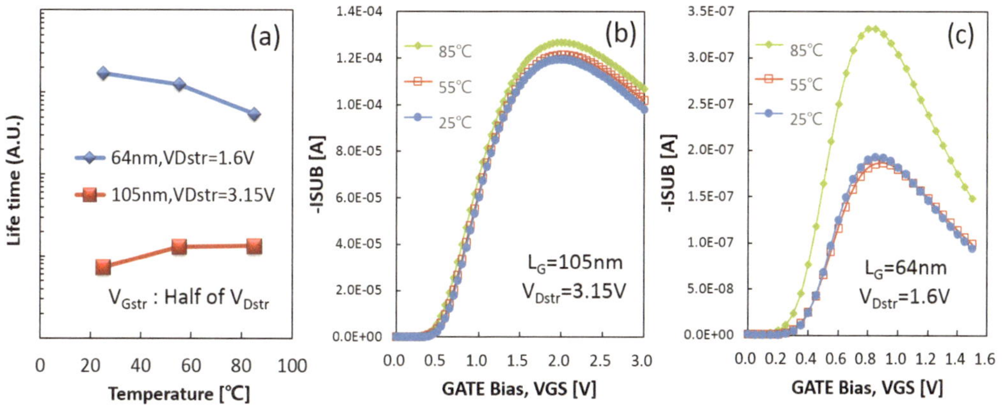

Figure 4. Hot carrier characteristics of 64-nm-gate-long, 1.6-V V_{DS} driving NMOSFET (SL) compares with that of 105-nm-gate-long, 3.15-V V_{DS} driving NMOSFET (LH). (a) Hot carrier lifetimes of SL and LH show different thermal dependencies: HC Lifetime of LH increases with temperature in accordance with the LEM framework, but does not for SL. (b) V_{GS}-dependent substrate currents of LH slightly enhance with temperature. (c) A remarkable increase at high temperature (85°C) found in SL suggests a temperature activation of HCD, which cannot be explained through the LEM framework.

region of the drain side. In LH transistors, hot carriers generated in the main region are dominant because of the large value of V_{DS}, which shows the negative dependency on temperature through λ. The scaled-down drain biases in short channel transistors reduce the V_{EFF} and the dominant hot carrier generation region is shifted from the main to the thermal tail via EES or MVHR. Since both are temperature-activating processes, the overall HCD shows a positive dependency on temperature that is detrimental especially in high-current-driving self-heating transistors like FinFETs.

2.3. PMOS hot carrier degradations

Traditionally, HCDs in PMOSFETs have not been taken seriously because a large energy barrier between Si and SiO_2 (~4.8 eV [24]) and a high-impact ionization threshold (Φ_{ii} = 1.43–1.92 eV [25]) of holes make difficult a LEM-like HCD in normal operational voltage ranges. The drain-avalanche hot-electron (DAHE) generates favorable electron injections into SiO_2 in low V_{GS} (1/3–1/4 of V_{DS}), which were known to be the dominant mechanism of HCD in PMOSFETs. The injected electrons fill the preexisting traps in the vicinity of the drain, which may cause effective gate length shortening, as $L_{eff} = L_o - \Delta L(N_{ox,e})$, and therefore punchthrough and breakdown may occur. However, this is a self-limiting procedure due to the exponential decrease of the electron injection current as a function of distance to the drain and hence yields a logarithmic dependency of $\Delta L(N_{ox,e})$ on time [26]. As PMOS gate oxide scales down, a turn-around of drain current degradation is observed, which is due to the charge re-emission and donor-like interface trap, $\Delta N_{it,d}$, generation under the high vertical field [25]. The dominant degradation driver has also been changed from the hot electron injection to the hot hole injection as Tox scales down. These transitions rely on (1) nitridation of gate oxide to suppress boron penetration, which enhances the generation of the positive charge (PC), (2) as oxide e-field, F_{ox}, exceeds 5MV/cm, NBTI degradation is triggered by cold hole injections at the source region, which are combined with hot hole injections at the drain region.

Abnormally large degradations of PMOS were reported in hot electron injection stress experiments at the cryogenic temperature of 77 K and subsequent anneal at elevated temperatures (300 K or higher) [27]. It is believed that the increase of carrier mobility and mean free path at 77 K creates additional damage sites in the oxide, which are initially inactivated at 77 K, and eventually convert to positive donor-type interface states as the de-trapped electrons leave vacancies in the annealing stage.

To summarize the hot carrier degradations in PMOSFETs, both electrons and holes created by impact ionization are responsible via their own natures for creating and/or changing the state of the oxide bulk traps and the interfacial traps. Since hot carrier generations in PMOSFETs are still negligible due to their low efficiency compared with those in NMOSFETs, cold hole injections to the SiO_2 can activate an appreciable number of interfacial and oxide bulk traps in the normal operational voltage range because holes are more efficient in trap generation processes than electrons are [28]. Cold hole injection is regarded as the most serious degradation mechanism of modern PMOSFETs. This subject is dealt with in Section 3 more precisely.

2.4. An energy-driven HCD modeling of NMOSFETs for circuit simulations

2.4.1. Aging model parameters

Transistors' degradation and the circuit performance degradation can be quantitatively analyzed through the circuit simulations by using the specific spice model parameter set, which we call "**aging parameters.**" Properly chosen aging parameters among the whole spice model parameters should be accurate over the full V_{DD} range varying V_{GS} and V_{DS} as a function of the "**age,**" which is an amount of "degradation." In summary, an age is accumulated during a prescribed operation time per transistor, the age shifts the aging parameters, and finally aging parameters reproduce the degraded transistors' characteristics. All the calculations are fulfilled during aging circuit simulations with self-consistent aging-parameter updates. A recursive process (age determines the degradation of transistor and vice versa) executes during the simulation. The complete sequence of the aging-parameters extraction and aging circuit simulations is schematically illustrated in **Figure 5**. Since the aging-parameters extraction is carried out under DC-stress conditions, some assumptions must be made regarding the validity of accumulated age and aging parameters updated during the AC circuit simulation, which include the following: (1) the static degradation rate and bias dependencies under DC stress conditions are assumed to be the same under AC stress conditions. This quasi-static approximation is generally accepted in HCD because the recovery after stress degradation is negligible and the total amount of degradation can be regarded as a singular function of AGE without any path dependencies; (2) the degradation is assumed to be a very slow process within the conventional time span of circuit simulation. It is an indispensable assumption for the sake of convenience and for the efficiency of the aging circuit simulation. It enables a decoupling of the aging accumulation from the aging-parameter update. One can accumulate age by using the voltage and current waveform, which is simulated with "constant" aging parameters during a prescribed time period, t_{CYC}. To control these non-overlapped sequences, aging circuit simulations can use two time variables, t and t_{AGE}. Age accumulation during t_{CYC} with time-invariant aging parameters is controlled by t. Aging parameters are subsequently updated by

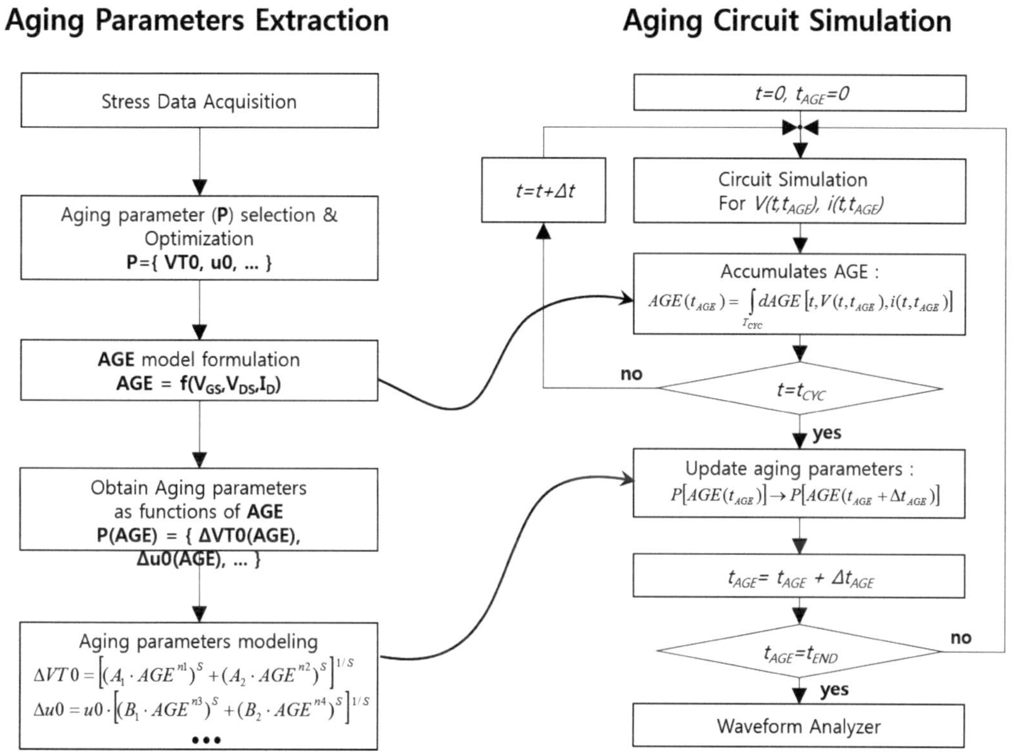

Figure 5. Schematics for aging-parameters extraction and aging circuit simulation procedures: Aging parameters (P) are selected from the transistor model parameters set to represent the experimental degradation with relevant physical mechanisms (*see left*). A unified degradation is modelled by an AGE and in turn aging parameters are extracted as a functionality of AGE. The aging circuit simulation is comprised of two non-overlapping loops, which are controlled by t and t_{AGE}, respectively (*see right*). The inner loop accumulates AGE with varying $V_{GS}(t)$ and $V_{DS}(t)$ for each transistor, and the outer loop updates the aging parameters using the accumulated AGE.

using the accumulated age as functions of $\mathbf{t_{AGE}}$. A flowchart depicted in **Figure 5** (right) illustrates the sequence of the aging circuit simulation in detail.

The proper sequence for the aging-parameter extraction can be exemplified in the following example: a 62-nm-gate-long NMOSFET is DC stressed with a V_{DS} within the range of 2.1–2.3 V and a V_{GS} within the range of 1.5–2.3 V. After a 300-s stress, transistors $I_D - V_{GS}$, and $I_D - V_{DS}$ are typically compared to a fresh one as in **Figure 6**.

A threshold voltage shift and transconductance, Gm reduction, are found in stressed $I_D - V_{GS}$ and Gm $- V_{GS}$ as shown in **Figure 6(a)**, and **(b)**. Selecting the spice model parameters **VTH0** for threshold voltage shift and ***u*0** for Gm reduction is the obvious choice for the aging parameters since $Gm = \frac{W}{L} C_{ox} V_{DS} \frac{\partial \mu}{\partial V_{GS}}$ in low V_{DS} and only $\frac{\partial \mu}{\partial V_{GS}}$ term can be degraded by hot carriers. As a coefficient of the mobility model, ***u*0** can scale both μ and $\frac{\partial \mu}{\partial V_{GS}}$ as shown in Eq. (6). One can find another important feature of degradation in **Figure 6(b)**; the reduction of the Gm-declining rate with V_{GS} is distinct. It is related to the increase of interface trap charges. They do screen more e-fields from the gate, hence the influence of the gate is reduced and the

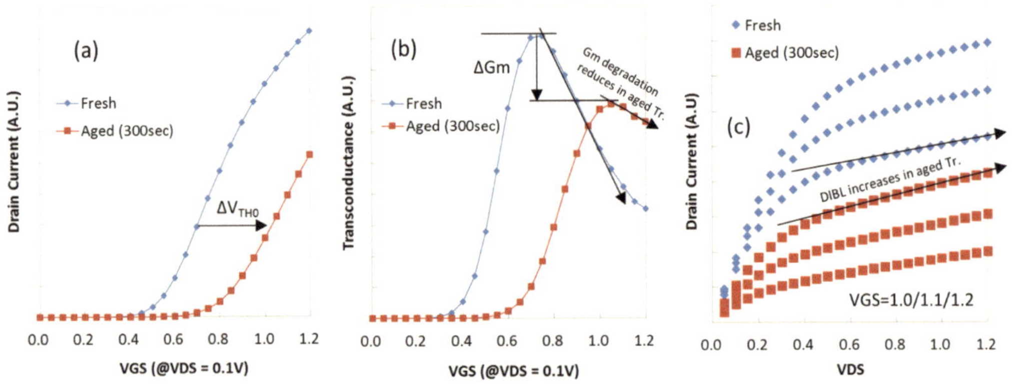

Figure 6. Comparisons of I_D-V_{GS}, Gm-V_{GS}, and I_D-V_{DS} showing the fresh and 300-s stressed with $V_{GS} = V_{DS} = 2.3$ V of 62-nm-gate-long transistors. (a) A comparison of fresh and aged transistors' I_D-V_{GS} clearly shows a stress induced **VTH0** shift. (b) Both the Gm and the Gm-declining rate on V_{GS} are reduced after being stressed. (c) A growing slope for the drain current on the V_{DS} is clearly shown in stressed I_D-V_{DS}'s, which can be interpreted as a DIBL increase due to HC stress.

surface-roughness-scattering-controlled Gm is less decreased in high V_{GS}. According to the spice model parameter equations, the Gm-declining rate can be modeled in the effective mobility, μ_{eff}, expressed in Ref. [29] as

$$\mu_{eff} = \frac{u0 \cdot f(L_{eff})}{1 + UA\left(\frac{V_{gsteff}+2V_{th}}{TOXE}\right) + UB\left(\frac{V_{gsteff}+2V_{th}}{TOXE}\right)^2 + UD\left(\frac{V_{th} \cdot TOXE}{V_{gsteff}+2\sqrt{V_{th}^2+0.0001}}\right)^2} . \quad (6)$$

UA or UB may adjust the declining rate on V_{GS} (V_{gsteff} in Eq. (6)). But it is not preferable as both ***u*0** and UA (or UB) appear in the same model equation, which makes it difficult to extract their optimum values independently. In other words, a lack of orthogonality may affect the quality of the parameter extractions. Thus, an alternative choice can be **rdsw**, which is a spice model parameter expressing the extrinsic resistance of drain and source regions. The drain resistance is increased by the accumulation of trapped electrons in the drain region. The Gm-declining rate is also affected by the accumulation of trapped electrons, which screen the gate electric field. Thus, choosing **rdsw** can include both a drain resistance increase and a Gm-declining rate decrease of degraded transistors without any ambiguity among parameters. The last parameter can be determined by observing **Figure 5(c)**. As the drain current-increasing slope along the V_{DS} is clearly shown in the aged transistors, one can choose a DIBL (drain-induced barrier-lowering) control parameter. An increase of DIBL originates from the same mechanism, which causes the increase of **rdsw**; the vertical e-field is screened by trapped charges and hence the channel inversion charges become more susceptible to the lateral e-field or V_{DS}. The DIBL formulation in the spice modeling [29] is

$$\Delta V_{th}(DIBL) = -\theta_{th}(DIBL) \cdot (ETA0 + ETAB \cdot V_{BS}) \cdot V_{DS} . \quad (7)$$

In Eq. (7), **ETA0** is a suitable parameter to describe the hot carrier-induced DIBL increase. Note that even though **VTH0** and **ETA0** may appear in the same threshold voltage model, they can be distinguished from each other since **VTH0** is extracted from HC degradation data without any dependency on V_{DS}, but **ETA0** is the coefficient of V_{DS} in ΔV_{th} that implies that one can extract both **VTH0** and **ETA0** independently. Selected aging parameters, **VTH0**, **u0**, **rdsw**, and **ETA0,** are optimized via appropriate numerical processes to best fit the experimental data. **Figure 7** compares the results where points mark the experimental data and lines are spice simulation results using optimized aging parameters.

2.4.2. An energy-driven AGE model

The AGE is a commonly used parameter to accumulate the amount of degradation under various bias conditions in aging circuit simulation. Appropriate AGE model reflects underlying physics with a relevant functional form for the bias and time. According to field-driven HCD, one can define the AGE function as

$$AGE_{\mathrm{FD}} = \frac{I_D}{W \cdot H}\left(\frac{I_{\mathrm{SUB}}}{I_D}\right)^m \cdot t_{\mathrm{STR}} \tag{8}$$

where m is known to be around 3 and H is a constant according to the field-driven HCD framework. **Table 1** checks the validity of this assumption. In this table, we can find that the largest VTH0 degradation occurs when $V_{DS} = V_{GS}$, among the various V_{DS}/V_{GS} bias sets, while the maximum substrate currents and AGEs do not coincide with that of VTH0 degradation. This mismatch implies that the field-driven mechanism is no longer valid for the 62-nm-scaled NMOS transistor.

In order to adjust the discrepancy of the field-driven AGE, one can modify H to be a function of V_{DG}, as a commonly used relief in the field. The fitting results will be compared with newly developed energy-driven AGE's results later. A simplified version of the energy-driven AGE model is proposed in Ref. [30] as

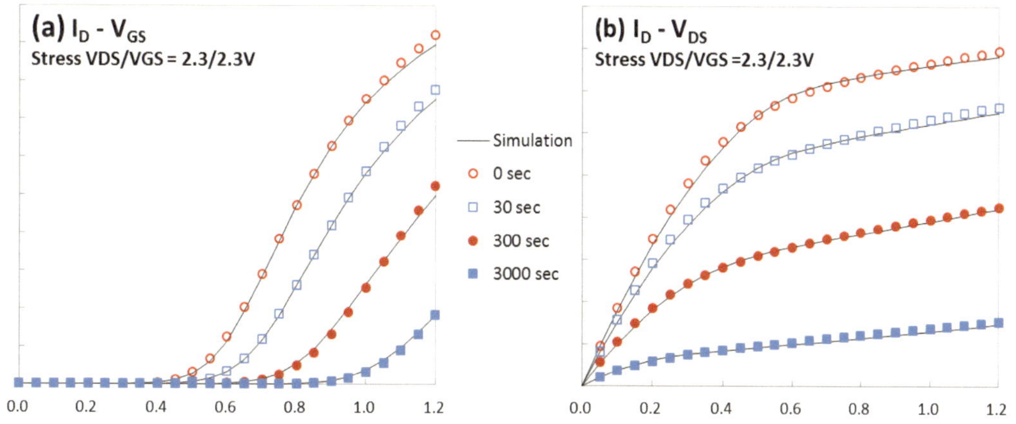

Figure 7. The aging-parameters optimization results fit the HCD measurements (points) with simulations (line) in (a) I_D-V_{GS} and (b) I_D-V_{DS} curves.

V_{DS}/V_{GS}	V_{DG}	I_{SUB}/I_D (x10^{-3})	AGE_FD (H = 1, m = 3)	Measured ΔVTH0 [mV]
2.1/2.1	0.0	0.491	3.32E-09	**151**
2.2/2.2		0.861	1.72E-08	**252**
2.3/2.3		1.572	9.97E-08	**395**
2.1/1.7	0.4	0.730	8.76E-09	76
2.2/1.8		1.210	3.97E-08	146
2.3/1.9		1.727	**1.01E-07**	257
2.1/1.3	0.8	**1.034**	**1.48E-08**	16
2.2/1.4		**1.754**	**8.44E-08**	95
2.3/1.5		**1.894**	8.61E-08	129

The maximum ΔVTH0 occurs at V_{DS} = V_{GS} (V_{DG} = 0V) for each V_{DS}. While I_{SUB}/I_D and AGEs do not follow ΔVTH0 tendencies on bias.

Table 1. Field-driven ages (Eq. (8)) are calculated from the measured I_{SUB}, I_D, and **VTH0** degradation (**ΔVTH0**).

$$R_{age} = a(V_G - V_{th})^P \cdot e^{B(V_D - V_{DSAT})} + b(V_G)^C \tag{9}$$

where R_{age} is the accumulation rate of the age, which is expressed by the multiplication of carrier density, $a(V_G - V_{th})^P$, carrier energy, $e^{B(V_D - V_{DSAT})}$ and $b(V_G)^C$ term for a high V_G dependency. Compared with Eq. (8), the linear carrier density dependency of I_D is generalized in Eq. (9) as having a power-law dependency with exponent P, which reflects the relevant mechanism of HC generation: one for field-driven, two for EES, and 20 for MVHR for a wide range of gate lengths of MOSFETs' and drain bias. The exponential term for carrier energy reflects the energy distribution of the electrons as a function of drain overdriving voltage, $V_D - V_{DSAT}$. The last term is negligible since it becomes significant only if the V_G is larger than 3.0, which is beyond the normal operation range in modern technology. The saturation voltage, V_{DSAT}, is originally defined by the drain current saturation point in MOSFETs' $I_D - V_{DS}$ relations. The velocity saturation of mobile carriers causes the drain current saturation of the scaled MOSFETs. At the same time, V_{DSAT} in the carrier energy distribution function defines the threshold energy to HCD. Although the notation V_{DSAT} is commonly used to denote the two different mechanisms, the values of V_{DSAT} for both mechanisms need not be the same. This is shown in Ref. [31] that the substrate current starts from a smaller V_{DSAT} but is still proportional to V_{DSAT}; thus the drain voltage dependency has the form of $V_D - \eta\, V_{DSAT}$, where η is a fitting constant defined within 0-1. From this observation, V_{DSAT}, as the threshold of HCD, should be extracted from both drain current and substrate current measurements, as a form of V_{DSAT}, or alternatively a new energy-driven AGE can be modified as

$$AGE_{ED} = (V_G - V_{th})^{n_G} \cdot \left(\frac{K}{\ln(I_D/I_{SUB})}\right)^{n_D} \cdot t_{STR} \tag{10}$$

where the well-known $I_{SUB}/I_D = C(V_D - \eta V_{DSAT})e^{-V_o/(V_D - \eta V_{DSAT})}$ form is used to replace $V_D - \eta V_{DSAT}$ by its simplified expression: $V_D - \eta V_{DSAT} \approx K/ln(I_D/I_{SUB})$ and the exponential function for the energy dependency is replaced by a power-law function because it has a better degree of freedom to fit to the experimental data. The last term of Eq. (9) is omitted as stated above. **Figure 8** compares the fitting results of the **ΔVTH0** by using the conventional e-field-driven AGE, Eq. (8), and the newly defined energy-driven AGE, Eq. (10). The *H* parameter in Eq. (8) is modified to have an exponential functional dependency of V_{GD} to fit the measurement data. As shown in the figure, the **ΔVTH0** tends to slow down as AGE increases. The saturation phenomenon is commonly found in the aging-parameter measurements. The interpretation for the saturation is that the current is pushed down by the interface electrons at the lightly doped drain region, which reduces the interface trap influence on the drain current reduction [32], or a saturation of preexisting charge trapping [33] results in a two-slope shape on the aging parameters dependent on the AGE. A behavioral expression of this effect can be generalized in the following expression:

$$\Delta P(\mathrm{AGE}) = \left[(P_1 \cdot \mathrm{AGE}^{n1})^S + (P_2 \cdot \mathrm{AGE}^{n2})^S\right]^{1/S} \tag{11}$$

where S is the shape factor and has a negative digit. The two-slope combination of Eq. (11) is used to fit the **ΔVTH0** dependence on the AGE as shown in **Figure 8(b)**. As shown in the figure, the overall consistency is improved by using the energy-driven AGE defined by Eq. (10). **Figure 9** illustrates the aging-parameters extraction results by comparing the measurement and the aging simulation results. Two kinds of extraction methods are compared in the graph, which are as follows: (1) AGEs are extracted by using only fresh measurement values of V_{th}, I_D, and I_{SUB} and (2) AGEs are extracted by degraded V_{th}, I_D, and I_{SUB} in order to reflect "degradation of age" recursively. The overall matching property is improved by this update as shown in the figure.

2.5. A hot carrier-resistant design technique through V_{GS} ratio controls

2.5.1. V_{GS} ratio and ADF

The last example is to demonstrate a hot carrier-resistant design technique. HC-resistant design techniques have been attracting more attention as technology gets smaller. A strong demand can be found in typical DRAM word-line driver circuits where the inherent risk of

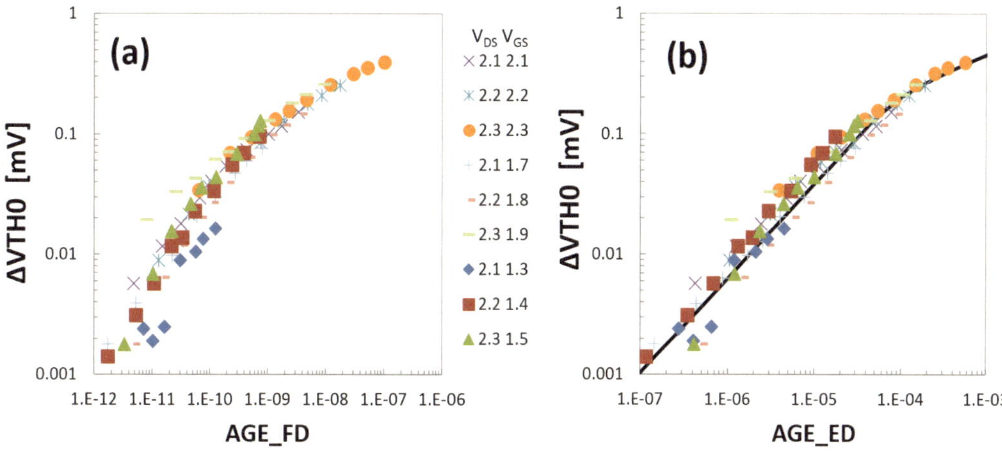

Figure 8. Optimized **ΔVTH0s** as a function of (a) the conventional field-driven age model (AGE_FD) and (b) the newly developed energy-driven age model (AGE_ED) are illustrated.

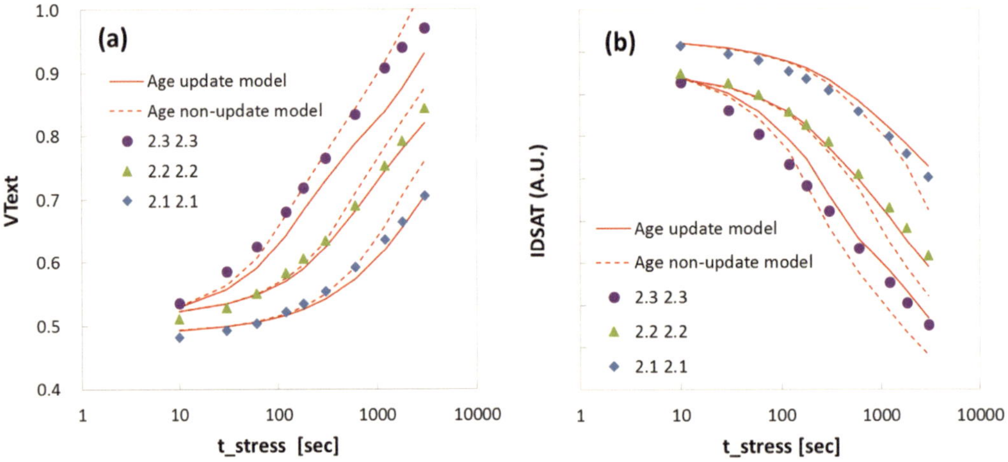

Figure 9. Aging-parameters' model extraction results are compared with the measurement data. As the age accumulation tends to slow down with the degrading ingredients (drain current and substrate current), the age update model parameter simulation results in (a) VText and (b) I_{DSAT} (solid lines) are less than those of the age non-update model (broken line). The age non-update model simulation results show an over-estimation especially in the deep degradation regime (t_{STRESS} > 1000 s). From this aspect, one can conclude that the overall accuracy of the simulation is improved by age update feedback.

HCD exists due to non-scale-down word-line-pumping voltage (V_{PP}). As stated above, the necessity for sustaining channel conductance in scaled cell transistors forces the V_{PP} to fix around 3 V. The field-driven mode can be a dominant HCD mechanism in such a high V_{PP}-biased 100-nm-long gate length transistors. According to the LEM, the maximum degradation occurs at the peak substrate current (I_{SUB}) generating V_{GS} condition. The peak I_{SUB} generation V_{GS} defines the "V_{GS} ratio", γ, which is $V_{GS,\,peak}/V_{DD}$. If the constant V_{DS} is applied under DC bias conditions, γ has its maximum value of around 0.5–0.6. The minimum value might be 1/3 since most CMOS transistors have their threshold voltage of 1/3 of V_{DD} and the drain voltage is assumed to pull down immediately as the gate turns on in the CMOS inverter operation. **Figure 10(a)** shows the HCD measurements with different V_{GS} ratios. The DC HCD reaches up to 40% at 100-h stress with γ = 0.5–0.6. Such a severe degradation does not seem to guarantee the lifelong serenity of the circuits without any kind of mitigation strategies. Several significant features of HCD are found in the figure:

1. As two time-slope phenomena are found in the figure as stated above, a sufficient timing margin is required to survive the initial rapid degradation. The overall timing shift of the WL driver reflects the "quasi-saturation effect" of the transistors' degradations as depicted in **Figure 10(b)**. As one can find in the figure, WL-off-degradation progresses toward a saturation at 5 years and very slowly degrades during further aging. In this design, wear-out is remarkably retarded due to this quasi-saturation phenomenon.

2. The V_{GS} ratio determines the quasi-saturation level of degradation. Compared at 100-h stress, only 1/3–1/2 of all degradations are shown in the γ = 1/3 stress condition as opposed to in γ = 0.5–0.6. Thus, the reduction of γ is the primary design target for long-term HCD reliability.

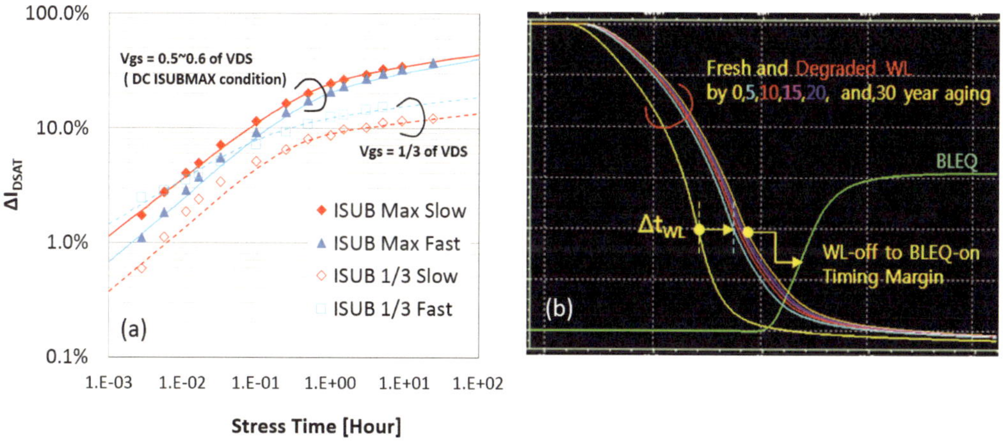

Figure 10. (a) I_{DSAT} degradations by HCI are depicted as functions of stress time. Two different choices of V_{GS} ratio affect amount of degradation and its quasi-saturation levels. The conventional choice of V_{GS} to let the substrate currents have their maximum among the V_{GS} values yields the maximum stress degradation of I_{DSAT} as shown in the figure. (b) The transistors' quasi-saturations influence on the word-line driver pull-down signal delays. The aging simulation by using the aging parameters reveals rapid degradations of word-line pull-down delay in the initial stress period (from 0 to 5 years) and their slowing down due to quasi-saturation in the subsequent period (from 5 to 30 years). It suggests that hot carrier-resistant design requires a sufficient timing margin to survive initial rapid degradation before the quasi-saturation takes place.

3. Process skews, which are mainly V_{TH} variations, affect HCD in a complicated manner. In the large γ case, a lower V_{TH} wafer (fast skew) has a smaller degradation than that of higher V_{TH} wafers (slow skew), and reverse for the small γ case. These phenomena can be explained by the two competing processes: (i) more gate overdriving, $V_{GS} - V_{TH}$ with lower V_{TH} reduces the lateral e-field and hence HCD, which predominates in the large γ case, (ii) lower V_{TH} increases the drain current as more electrons participate in the impact ionization process and hence cause higher HCD, which predominates in the small γ case. Thus, V_{TH} controls in order to improve HCD may cause the opposite results depending on γ.

As shown in **Figure 10**, we can conclude that the V_{GS} ratio has an effect not only on degradation rate but also on its quasi-saturation level by ×2–×3 differences. The quasi-saturation level of the degradation is especially important in long-term HCI degradation where most transistors suffer sufficient stress to enter the quasi-saturation region.

AC duty factor (ADF) is commonly used to estimate the HCD in AC operations. It is defined by the AGE in DC bias conditions divided by the aggregation of the AGE per circuit operation cycle. The commonly used form of ADF is shown as

$$\mathrm{ADF} = \frac{I_{SUB\_DC}{}^{m} \cdot t_{RC}}{\int_0^{t_{RC}} I_{SUB}(t)^m \cdot dt} \quad . \tag{12}$$

As a large value for ADF improves AC hot carrier reliability, a straightforward HCI-resistant design may focus on as large a value as possible. However, this is not a necessary condition for long-term HCI-resistant design. **Figure 11** illustrates the HCD of critical transistors consisted

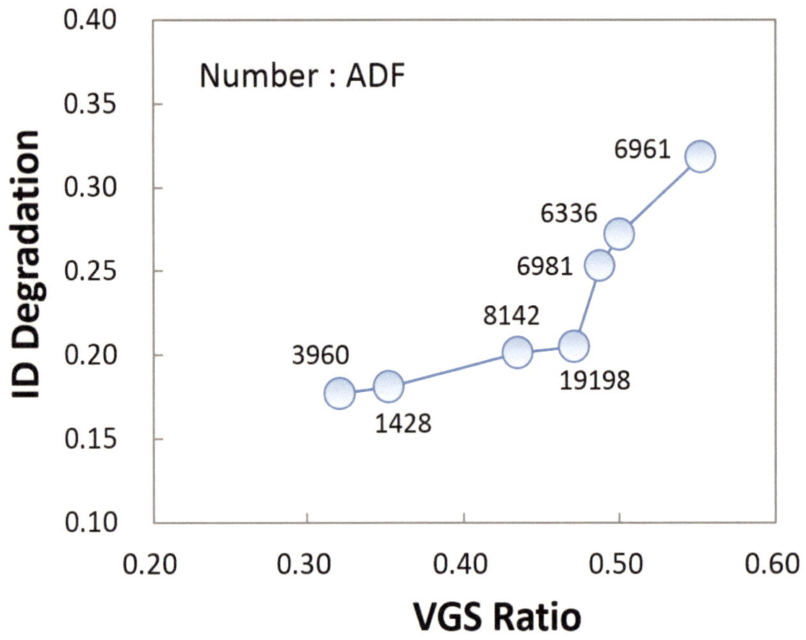

Figure 11. Transistor's current degradations are simulated to reproduce transistor AC degradation after a 10-year operation of a WL driver circuit. Simulation results are depicted with respect to the V_{GS} ratio. AC duty factors (ADFs) are also marked in the same graph as numbers.

of the word-line driver circuitry after a 10-year aging period, which is long enough to enter the quasi-saturation regime for all transistors. The significant role of the V_{GS} ratio can be reconfirmed in the figure. ADF may retard degradation while it has no influence on ΔI_D once it enters the saturation region. From this observation, we can conclude that reducing the V_{GS} ratio has a significant effect on the long-term hot carrier reliability. By contrast, a large ADF retards the HCD to reach its quasi-saturation level, but has no effect on reducing the quasi-saturation level itself.

Driver strength is the most important control parameter regarding circuit level HCI degradation. Strong drivers can easily pull down the output voltage, V_{DS}. Due to the exponential dependency on V_{DS}, the substrate current quickly diminishes with the fast pull-down and, as a consequence, ADF increases and γ decreases, respectively. **Figure 12** illustrates a typical example of substrate current shape as a function of V_{GS} in a CMOS inverter circuit. By increasing driver strength, the substrate current peak is decreased and hence HCI stress is mitigated.

2.5.2. HCI-resistant design strategy

As shown above, reducing the V_{GS} ratio and increasing ADF can be recommended as HCI-resistant design strategies. Reducing input slew rate and increasing output slew rate or increasing driver size is a straightforward method to obtain both design targets. Reducing

V_{DS} humps through increasing gate-drain overlap capacitance of the strong driver is an additional benefit to mitigate HCD. One can find an example of HCI-resistant design for a typical inverter chain logic depicted in the inset of **Figure 13(a)**. As driver size splits from ×1 to ×2, ×4, and ×8, input and output slew rates are correctly modified and the I_{SUB} waveforms go

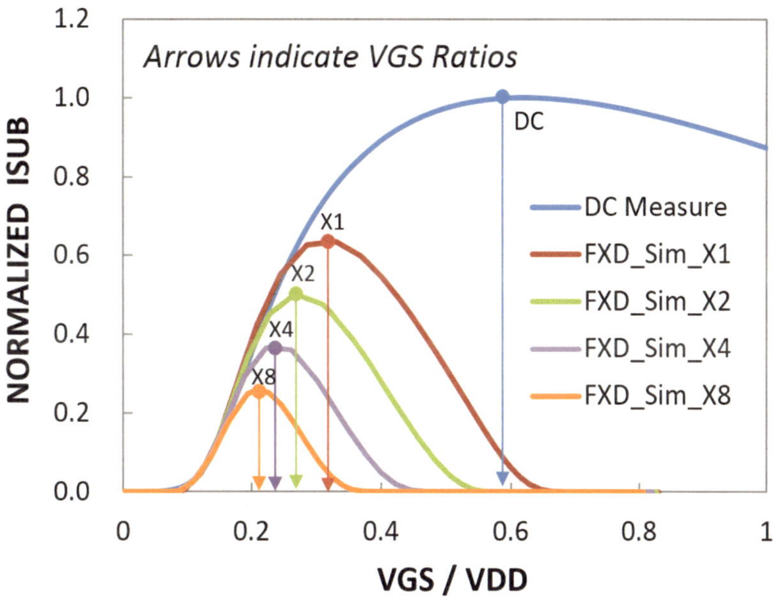

Figure 12. Substrate current profiles are compared as functions of normalized gate bias (V_{GS}/V_{DD}) in various driver strengths. The V_{GS} ratio is maximized in DC sweep mode (constant V_{DS} is applied) and decreased in CMOS inverter operation mode (increasing V_{GS} pulls down V_{DS}). The driver strength strongly influences both substrate currents and the V_{GS} ratios. From this context, one can deduce that the long-term hot carrier reliability can be improved with driver strength.

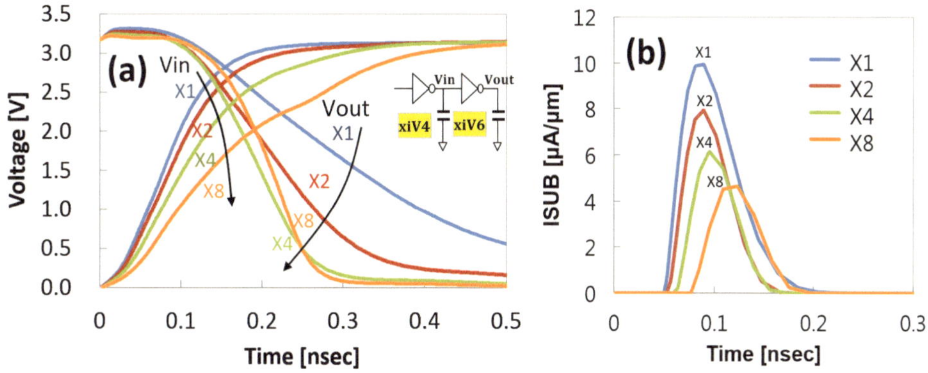

Figure 13. (a) Simulated input and output voltage waveforms of an inverter driver (xiV6), which are measured at the gate node and the drain node of xiV6, respectively. Input slews decrease and output slews increase with driver strength results in deductions in I_{SUB} as depicted in (b).

down as shown in **Figure 13(b)**. The HCD simulation for the inverter chain consisting of pre-driver (xiV4) and main-driver (xiV6) results is depicted in **Figure 14** with main driver size splits. The increase in size reduces HCD of the main driver as shown in the figure with a slight loss of that of the pre-driver. The increase in driver size, or fan-out, means more layout area consumption and a deterioration of pre-driver's HCD, and hence should be compromised. A trade-off design choice between HCI robustness and area penalty may exist within the range of ×2–×4 in the examples.

2.6. Summary

The hot carrier degradation mechanism has evolved from the single-particle e-field-driven model to the multi-particle energy or current-driven model associated with consistent technology scale-downs. During the last 30 years, a general agreement has been made that the complete extinguishing of this kind of catastrophic failure is impossible, though a better understanding of the physical mechanism and the relevant modeling contributes to developing HCI-aware design techniques. As a part of which the aging-parameters optimization of HCD of a 62-nm gate-long NMOS transistor has been demonstrated. Newly developed energy-driven AGE formulations show a better consistency with the experimental data than conventional e-field-driven AGE models without any arbitrary correction function of H, which implies that an energy-driven HCD predominates in the deca-nanometer-scaled transistors. A hot carrier-resistant design example for the DRAM word-line driver is also presented. From this study, a long-term hot carrier-resistant design strategy can be summarized as follows: (1) Give sufficient timing margins to survive the rapid initial degradation. (2) The V_{GS} ratio is the key control parameter since it directly relates to the quasi-saturation level of long-term HCD. (3) A prime design target is driver strength, or fan-out. Stronger drivers reduce the HCD through the increase of ADF and, more importantly, through the decrease of the V_{GS} ratio but at the cost of an area penalty. A compromise between driver strength and area penalty is a required trade-off in HCI-resistant design solutions.

Figure 14. (a) Fan-out increasing enhances V_{GS} ratio and ADF in the xiV6 driver while both HCI-related parameters (V_{GS} ratio and ADF) deteriorate in the XiV4 pre-driver. (b) I_D degradation after a 10-year operation increases for xiV6, as opposed to decreases for xiV4. An optimal trade-off of driver strength (fan-out) considering both long-term HCI resistance and area penalty can be made in x2–x4 range of driver strength (fan-out).

3. Negative bias temperature instabilities (NBTI) in PMOSFETs

3.1. Introduction

As deca-nanometer-scaled transistors require the oxide scaled down to around 20 Å for sufficient gate control, a 5–10 MeV/cm of vertical e-field is easily established in the nitride-cooperated silicon dioxide region under the normal V_{DD} conditions. The Fowler-Nordheim tunneling mechanism can be triggered in such a high e-field, negative bias applied in PMOS gates collect inversion holes and then tunneling into gate oxide by e-field driving. Although this range of e-field is not sufficiently strong to generate hot carriers, cold holes in PMOS can interact with the oxide bulk traps and hydrogen passivated in the Si/SiO_2 interface with temperature activation, which in turn result in positive charges in the oxide region and the Si/SiO_2 interface. The critical characteristics of PMOS, threshold voltage (V_T), drain current (I_D), and transconductance (Gm), can be degraded by the trapped positive charges once the gate is negatively biased in moderate thermal conditions, regardless of V_{DS} or drain current. Owing to its nature, negative bias temperature instability (NBTI) has been an urgent issue in state-of-the-art PMOS transistors, which are prone to gate-tunneling hole-induced degradation. NBTI-resistant design is quite difficult because a simple turn-on operation triggers NBTI degradation. This means that the degradation occurs during the whole period of PMOS turn-on, and thus the only possible way to prevent it seems to be by a "power cut" during PMOS standby periods, which incurs spatial and performance penalties.

When the negative bias is applied, oxide bulk charges, which are called E′ centers (oxygen vacancies) interact with the holes or protons (H^+) to produce positive charge build-up in the oxide bulk [34]. Interface trap (N_{it}) generation is attributed to a breaking of Si–H bonds by holes [35], or by protons [36], which leave behind the amphoteric trivalent $Si_3 \equiv Si^{\bullet}$ defects, that is, silicon-dangling bonds, which are called P_b centers. Thermal nitride deposition during gate oxidation has been known to cause additional profiles of the positive trap states, which are called DLHT [9, 10] or positive charges (PCs) [37]. When the stress bias turns off, a part of the PCs can be neutralized by the bulk electron from the N-Well, which causes a quick recovery after stress. The NBTI-degrading species and related mechanisms can be summarized as follows:

1. Gate oxide-injected cold holes create positive charges through dissociation of Si–H bonds at the interface (ΔN_{it}) or by being captured in the E′ center in the oxide bulk, which is responsible for ΔN_{ot}.

2. Atomic hydrogen, H^+, or protons predominately supplied by end-of-line anneal steps can generate interface traps, N_{it} through the following chemical reaction [36]:

$$(\mathrm{Si-H}) + \mathrm{H}^+ \leftrightarrow (\mathrm{Si}^*)^+ + \mathrm{H}_2 \tag{13}$$

   where (Si*) denotes the unbonded silicon lattice (P_b center) and the remaining symbols have their conventional meanings.

3. Molecular hydrogen is responsible for the annealing of E′ traps through repassivation of one hydrogen atom from H_2, and the other diffuses away [34].

4. Positive charges (PCs) are generated during negative stress. Distinct increases of PCs are found in thermally nitrided oxide (TNO), which possibly originate from the emission of electrons from nitrogen donors [9]. A part of the PCs is quickly neutralized in the subsequent recovery cycle by electron injection from the N-Well.

Except for point (4), the interface trap charge generation, ΔN_{it} tends to be proportional to oxide bulk-charge generation, ΔN_{ot} [38, 39] since they commonly originate from the same species, hole and H^+. One may focus only on ΔN_{it} to understand the role of hydrogen upon NBTI. The positive charge generation process elucidated in point (4) can be neglected since it can be regarded as only a short-term process, which has little effect on the long-term reaction and diffusion of hydrogen. To show a practical case study in the subsequent section, one can draw a qualified image of NBTI through a reaction and diffusion model of hydrogen for a better understanding of NBTI-aware process integrations.

3.2. Reaction-diffusion framework

The generalized form of N_{it} rate equation is

$$\frac{dN_{\mathrm{it}}}{dt} = k_F(N_0 - \Delta N_{\mathrm{it}} - N_{\mathrm{it},0}) - k_R(\Delta N_{\mathrm{it}} + N_{\mathrm{it},0}) \cdot {H_0}^{1/\alpha} \tag{14}$$

where H_0 is the Si/SiO_2 interface concentration of hydrogen species ($\alpha = 1$ for atoms, 2 for molecules), and N_0 is the Si/SiO_2 interface concentration of Si–H bonds, $N_{\mathrm{it},0}$ is the preexisting interface trap density and ΔN_{it} is the newly generated interface trap density, respectively. The dominant species of diffusion into the oxide and gate region is known to be molecular hydrogen, which can be governed by the following two-dimensional diffusion equation:

$$\frac{\partial H}{\partial t} = D_H \frac{\partial^2 H}{\partial x^2} + D_H \frac{\partial^2 H}{\partial y^2}. \tag{15}$$

Since the N_{it} originates from the dissociation of hydrogen at the interface, the total amount of N_{it} is assumed to be equal to (1) the sum of the hydrogen species in the gate regions plus (2) the amount of hydrogen species diffused out from the gate regions. The numerical expression may be expressed as follows

$$N_{\mathrm{it}} = \int_{\text{Gate Regions}} H(x,y,t) \cdot dxdy + \int_{t'=0}^{t} \oint_{\text{bound}} k_p H(x,y,t') \cdot dS \cdot dt' , \tag{16}$$

with the assumption that all the hydrogen species, which reach the boundary are absorbed at a surface absorption velocity, k_p [cm/s] is expressed as

$$-D_H \cdot \nabla H|_{\text{boundary}} = k_P H|_{\text{boundary}} . \tag{17}$$

The existence of the ideal sink at the boundary is a rather unphysical assumption but its exclusion is unavoidable because it is impossible to remove the complicated ambient effect of the outside hydrogen. Furthermore, it improves the feasibility of the simulation. The diffusion constant of the hydrogen species, D_H, has strong material dependencies as shown in Ref. [7], which indicates that the diffusion speed of neutral species (H0, H2) is highest in the oxide, next in the poly, next in Si substrate, and extremely slow in the nitride film. Since hydrogen diffusivities have strong dependencies on material and are assumed to also have strong dependencies on film deposition conditions and even on the structure, one may rely on a numerical optimization method, as the only feasible way to determine each parameter (D_H, k_F, k_R, k_p, and N_0). Although physical uncertainty may be caused by a numerical optimization and it can be a major drawback for the reaction-diffusion framework, it is still valid for the relative analysis based on comparisons between the experimental data with consistent usage of optimized parameters. Simulation results are compared to the experimental value in **Figure 15**. A timing exponent measured as 0.17 for 10–1000 s as shown in the figure suggests that the diffusion species is molecular hydrogen [40]. An increase of N_{it} rate for 1000–10,000 s is ascribed to the out-diffusion of hydrogen through boundary layers. The N_{it} rate goes down and finally saturates after 100,000 s as N_{it} approaches N_0, which was also predicted in [40].

3.3. End-of-line anneal effects on NBTI

The preexisting hydrogen, $H(x, y, 0) = H_{PRE}$ seems not to have any influence on the N_{it} slope because the diffusion equation of hydrogen (Eq. (15)) is independent of N_{PRE}. However, atomic hydrogen also generates N_{it} as well as holes do as indicated in Eq. (13). One can postulate that

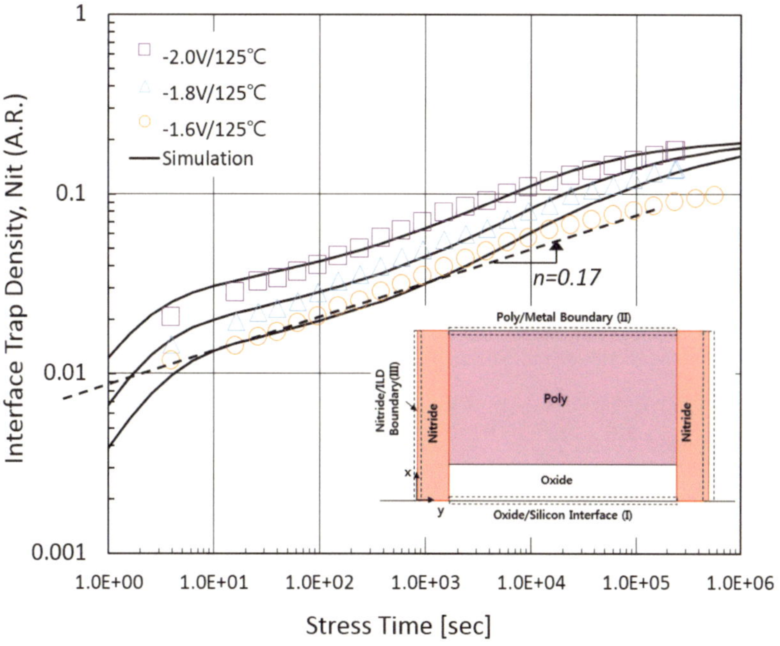

Figure 15. The simulation results of the reaction-diffusion of hydrogen are compared to measured NBTI degradations expressed as the interface trap generation rates. Inset: The simulation region consists of the oxide, poly, sidewall nitrides, and boundaries.

N_{PRE} can "increase" N_0 assuming that non-overlapped energy bands exist in the Si-H dissociation process caused by holes and by atomic hydrogen, respectively. Evidence for this assumption can be found in charge-pumping measurements and reproduced simulation results as illustrated in **Figure 16**. Interface trap densities, N_{it}, are measured by the charge-pumping method before and after 300-s-NBTI stress. The increase of N_{it} or ΔN_{it} after stress shows inversely proportional relations to the initial N_{it} or $N_{it,0}$ as shown in the figure. One can interpret this dependency through Eq. (14) as the forward reaction term decreases as an increase of $N_{it,0}$ and the reverse reaction (repassivation of the interface trap) term increases with $N_{it,0}$.

An intriguing aspect is found in parameter fitting for sample A, B1, and B2: using a higher value of N_0 is indispensable for sample A to fit the experimental ΔN_{it} rather than for samples B1 and B2. This difference is assumed to originate from different EOL anneal condition splits for samples A, B1, and B2. All the samples undergo the hydrogen passivation process in a very low-pressurized anneal chamber with N_2 ambient at temperatures of 365–390°C (B2 sample experiences higher temperatures than others). Additional forming gas (H_2 + N_2) anneal applies only in sample A in the atmospheric chamber with 390°C. The precedent anneal applied to all the samples can passivate the silicon-dangling bonds with hydrogen, which is believed to come from the hydrogen-rich passivation layer that covered all the wafers, and the remaining hydrogen species may diffuse out from the silicon due to the

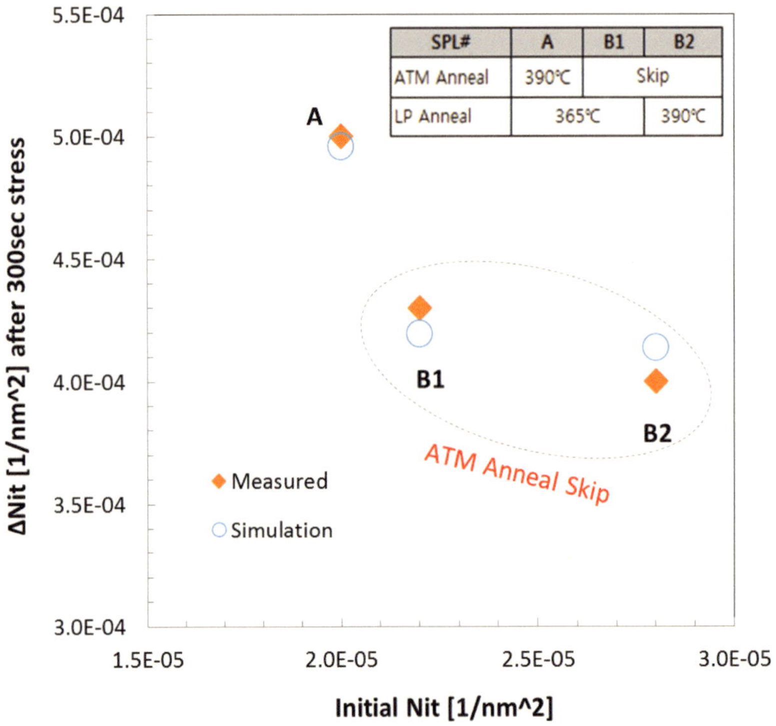

Figure 16. Charge-pumping measurement data before and after 300-s-NBTI-stress are compared to the R-D simulation results.

low-pressure ambience. Additional EOL anneal applies only to sample A, which can passivate additional silicon-dangling bonds and may leave a volume of hydrogen species in the gate regions. It causes additional passivation to reduce initial interface trap density, $N_{it,0}$ and to increase ΔN_{it} after NBTI stress by the remaining hydrogen, which can be expressed as the increase of N_0 in Eq. (14).

The evidence of remaining hydrogen, which is supposedly interstitial hydrogen, and induced N_0 enhancement, is also found in the anneal time split results shown in **Figure 17**. The low-pressure anneal time is doubled in the split group and it shows an earlier saturation of ΔN_{it} than the control group as shown in **Figure 17**. More interstitial hydrogen can diffuse out during the extended anneal step. It is believed that the reduction of interstitial hydrogen through the extended low-pressure anneal can reduce ΔN_{it} during NBTI stress. This is also reproduced in the simulation with two assumptions: a 25% increase in N_0 and the preexistence of hydrogen for the control group as compared in **Figure 17**. This conflict results in passivation anneal splits; ΔN_{it} increases with the additional atmospheric anneal, but decreases with the additional low-pressure anneal, which strongly suggests that the remaining hydrogen can make additional Si–H bond breakage plus that which the holes do. This can be reproduced by reaction-diffusion simulations through the increase of N_0 and H_{PRE}. From this plausible interpretation, one can conclude that removing hydrogen as much as possible from the transistor gate regions improves the long-term NBTI reliability.

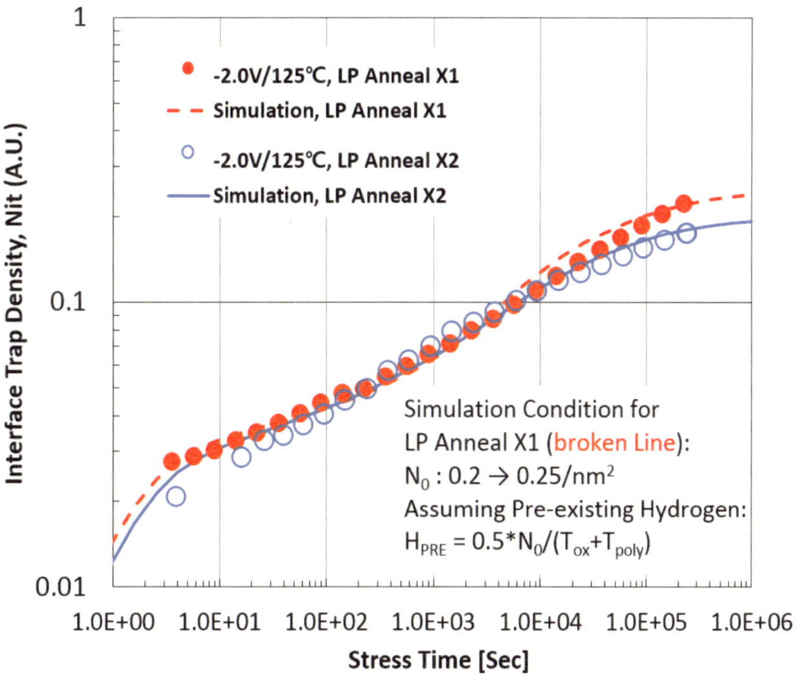

Figure 17. Interface trap generation rates of the low-pressure anneal split samples are compared to the R-D simulation results. Incomplete anneal split (LP Anneal X1 depicted in red line) is modelled by (1) increasing N_0 from 2.0 to 2.5 /nm^2 and (2) assuming pre-existing hydrogen in the gate regions.

3.4. Summary

NBTI has become the predominate long-term reliability threat as gate oxide is scaled down to the 20-Å range. It is caused by various source species: channel hole injections into gate oxide breaks the Si–H bond. Preexisting atomic hydrogen also dissociates Si–H bond to form a molecular hydrogen. Oxide bulk traps or nitride traps can be activated by holes or atomic hydrogen captures, or emitting electrons. Molecular hydrogen can neutralize the oxide bulk traps. Positive charges quickly generate and disappear at the initial stage of stress and recovery, respectively, which is believed to be associated with the nitrogen donor traps. The reaction-diffusion model simulation has demonstrated to reproduce the experimental ΔN_{it} data with the following features: a power-law functional dependency of 0.17 timing exponent for 10–1000 s and an eventual increase and saturation during the subsequent stress period. It proves that the reaction-diffusion framework of hydrogen, although still controversial among researchers, evidently reproduces NBTI degradation characteristics. Another R-D analysis of hydrogen is demonstrated in the EOL anneal experiments, which reveal that (1) preexisting hydrogen cooperates with holes to break the Si–H bonds, which can be modeled by an increase of the total number of breakable Si–H bonds, N_0. (2) Removing unbonded hydrogen from the transistor gate improves the long-term NBTI reliability.

Acknowledgements

The author would like to express his appreciation for the device modeling and reliability group and the DRAM device group of SK hynix for providing valuable measurement data and discussions for this chapter. The author would also like to acknowledge Lance S. Phipps who carefully reviewed and corrected the sentences in the manuscript to improve readability.

Author details

Chang Yeol Lee

Address all correspondence to: changyeol.lee@sk.com

SK hynix, Icheon, South Korea

References

[1] Intel Chips timeline. [Internet].2012. Available from: http://www.intel.com/content/www/us/en/history/history-intel-chips-timeline-poster.html [Accessed: 2017-05-08]

[2] Bornstein W, et al. Field degradation of memory components from hot carriers. In: IEEE 44th Annual International Reliability Physics Symposium (IRPS); San Jose; 2006. pp. 295-305

[3] Hu C, et al. Hot-electron-induced MOSFET degradation—Model, monitor, and improvement. IEEE Journal of Solid-Sate Circuits. 1985;**SC-20**(1):295-305

[4] El-Mansy YA, Boothroyd AR. A simple two-dimensional model for IGFET operation in the saturation region. IEEE Transactions on Electronic Devices. 1977;**ED-24**(3):254-262

[5] Hess K, et al. Giant isotope effect in hot electron degradation of metal oxide silicon devices. IEEE Transactions on Electronic Devices. 1998;**45**(2):406-416

[6] Li E, et al. Projecting lifetime of deep submicron MOSFETs. IEEE Transactions on Electronic Devices. 2001;**48**(4):671-678

[7] Krishnan AT, et al. Material dependence of hydrogen diffusion: Implications for NBTI degradation. In: International Electron Devices Meeting (IEDM); Washington, DC, USA, 2005. pp. 4-7

[8] Kufluoglu H, et al. A geometrical unification of the theories of NBTI and HCI time-exponents and its implications for ultra-scaled planar and surround-gate MOSFETs. In: International Electron Devices Meeting (IEDM); San Francisco, CA, USA, 2004. pp. 113-116

[9] Ang DS, et al. Consistent deep-level hole trapping model for negative bias temperature instability. IEEE Transactions on Device and Materials Reliability. 2008;**8**(1):22-34

[10] Ang DS, et al. Effect of hole-trap distribution on the power-law time exponent of NBTI. IEEE Electronic Device Letters. 2009;**30**(7):751-753

[11] Randriamihaja YM, et al. New hot carrier degradation modeling reconsidering the role of EES in ultra short N-channel MOSFETs. In: IEEE International Reliability Physics Symposium (IRPS); Monterey, CA, USA, 2013. pp. XT.1.1-XT.1.5

[12] Rauch SE, III, et al. Role of E-E scattering in the enhancement of channel hot carrier degradation of deep-submicron NMOSFETs at high VGS conditions. IEEE Transactions on Device and Materials Reliability. 2001;**1**(2):113-119

[13] Bravaix A, et al. Hot-carrier to cold-carrier device lifetime modeling with temperature for low power 40 nm Si-bulk NMOS and PMOS FETs. In: International Electron Devices Meeting (IEDM); Washington, DC, USA, 2011. pp. 622-625

[14] Guerin C, et al. The energy-driven hot-carrier degradation modes of nMOSFETs. IEEE Transactions on Device and Materials Reliability. 2007;**7**(2):225-235

[15] Suk SD, et al. Characteristics of sub 5nm tri-gate nanowire MOSFETs with single and poly Si channels in SOI structure. In: Symposium on VLSI Technology (SOVT); Kyoto, Japan, 2009. pp. 142-143

[16] Cho H-J, et al. Si FinFET based 10 nm technology with multi V_t gate stack for low power and high performance applications. In: Symposium on VLSI Technology (SOVT); Honolulu, HI, USA, 2016

[17] Guo D, et al. FINFET technology featuring high mobility SiGe channel for 10nm and beyond. In: Symposium on VLSI Technology (SOVT); Honolulu, HI, USA, 2016

[18] Khan MI, et al. Self-heating and reliability issues in FinFET and 3D ICs. International Conference on Solid-State and Integrated Circuit Technology (ICSICT); Guilin, China, 2014

[19] Miyashita T, et al. High voltage I/O FinFET device optimization for 16nm system-on-a-chip (SoC) technology. In: Symposium on VLSI Technology (SOVT); Kyoto, Japan, 2015. pp. T152-T153

[20] Jin M, et al. Reliability characterization of 10 nm FinFET technology with multi-VT gate stack for low power and high performance. In: International Electron Devices Meeting (IEDM); San Francisco, CA, USA. 2016. pp. 380-383

[21] National Chiao Tung University, Taiwan. Information Technology Service Center. National Chiao Tung University. Silicon Surface Density [Internet]. 2017. Available from: http://web.it.nctu.edu.tw/~thhou/09-dee4515/Silicon%20Surface%20Density.pdf [Accessed: 2017-05-08]

[22] Bude JD, et al. Impact ionization and distribution functions in sub-micron nMOSFET technologies. IEEE Electronic Device Letters. 1995;**16**(10):439-441

[23] Zanchetta S, et al. Analytical and numerical study of the impact of HALOs on short channel and hot carrier effects in scaled MOSFETs. Solid-State Electronics. 2002;**46**:429-434

[24] Hofmann KR, et al. Hot-electron and hole-emission effects in short n-channel MOSFET's. IEEE Transactions on Electronic Devices. 1985;**ED-32**(3):691-699

[25] Bravaix A, et al. Hole injection enhanced hot-carrier degradation in PMOSFETs used for systems on chip applications with 6.5-2 nm thick gate-oxides. Microelectronics Reliability. 2004;**44**:65-77

[26] Brox M, et al. A model for the time- and bias-dependence of p-MOSFET degradation. IEEE Transactions on Electronic Devices. 1994;**41**(7):1184-1196

[27] Song M, et al. Comparison of NMOS and PMOS hot carrier effects from 300 to 77K. IEEE Transactions on Electronic Devices. 1997;**44**(2):268-276

[28] Brox M, et al. Dynamic degradation in MOSFET's—Part I: The physical effects. IEEE Transactions on Electronic Devices. 1991;**38**(8):1852-1858

[29] Dunga MV, et al. BSIM4.6.0 MOSFET model – user's manual. Department of Electrical Engineering and Computer Sciences. University of California, Berkeley; 2006

[30] Arfaoui W, et al. Application of compact HCI model to prediction of Process effect in 28FDSOI technology. In: IEEE International Integrated Reliability Workshop (IIRW); S. Lake Tahoe, CA, USA, 2014. pp. 69-72

[31] Yang JJ, et al. A new approach to modeling the substrate current of pre-stressed and post-stressed MOSFET's. IEEE Transactions on Electronic Devices. 1995;**42**(6):1113-1119

[32] Dreesen R, et al. Modelling hot-carrier degradation of LDD NMOSFETs by using a high-resolution measurement technique. Microelectronics Reliability. 1999;**39**:785-790

[33] Lachenal D, et al. HCI degradation model based on the diffusion equation including the MVHR model. Microelectronic Engineering. 2007;**84**:1921-1924

[34] Tahanout C, et al. Oxide trap annealing by H2 cracking at E′ center under NBTI stress. In: International Conference on Microelectronics (ICM); Algiers, Algeria, 2012

[35] Fujieda S, et al. Interface defects responsible for negative-bias temperature instability in plasma nitride SiON/Si (100) systems. Applied Physics Letters. 2003;**82**(21):3677-3679

[36] Tsetseris L, et al. Physical mechanisms of negative-bias temperature instability. Applied Physics Letters. 2005;**86**:142103-142105

[37] Hatta SWM, et al. energy distribution of positive charges in gate dielectric: Probing technique and impacts of different defects. IEEE Transactions on Electronic Devices. 2013;**60**(5):1745-1753

[38] Huard V, et al. Evidence for hydrogen-related defects during NBTI stress in p-MOSFETs. In: International Reliability Physics Symposium (IRPS); Dallas, Texas, USA, 2003. pp. 178-182

[39] Tan SS, et al. Nitrogen-enhanced negative bias temperature instability: An insight by experiment and first-principle calculations. Applied Physics Letters. 2003;**82**(12):1881-1813

[40] Chakravarthi S, et al. A comprehensive framework for predictive modeling of negative bias temperature instability. In: International Reliability Physics Symposium (IRPS); 2004. pp. 273-282

Section 3

IC Design and Fabrication

Chapter 3

Low Power Design Methodology

Vithyalakshmi Natarajan, Ashok Kumar Nagarajan,
Nagarajan Pandian and Vinoth Gopi Savithri

Additional information is available at the end of the chapter

http://dx.doi.org/10.5772/intechopen.73729

Abstract

Due to widespread application of portable electronic devices and the evaluation of micro-electronic technology, power dissipation has become a critical parameter in low power VLSI circuit designs. In emerging VLSI technology, the circuit complexity and high speed imply significant increase in the power consumption. In low power CMOS VLSI circuits, the energy dissipation is caused by charging and discharging of internal node capacitances due to transition activity, which is one of the major factors that also affect the dynamic power dissipation. The reduction in power, area and the improvement of speed require optimization at all levels of design procedures. Here various design methodologies are discussed to achieve our required low power design concepts.

Keywords: power modeling, switching activity, self-transition, coupling transition, low power dissipation, VLSI

1. Introduction

As VLSI technology advances, the complexity and speed circuit increase, resulting in high power consumption. In VLSI design, small area and high performance are two conflicting constraints. The integrated circuit (IC) designer's activities have been involved in trading of these constraints. There are many possible design considerations, due to which the power efficiency has become important. The most portable systems used in recent era, which are powered by batteries, are performing tasks requiring lots of computations. The most important aspect of Moore's Law is that it has become a universal predictor for the growth of the entire semiconductor industry. From Moore's law, it is understood that the number of devices in a chip doubles every 18 months. This will increase the number of transistors used and hence increase the area and power consumption of the circuit (**Figure 1**).

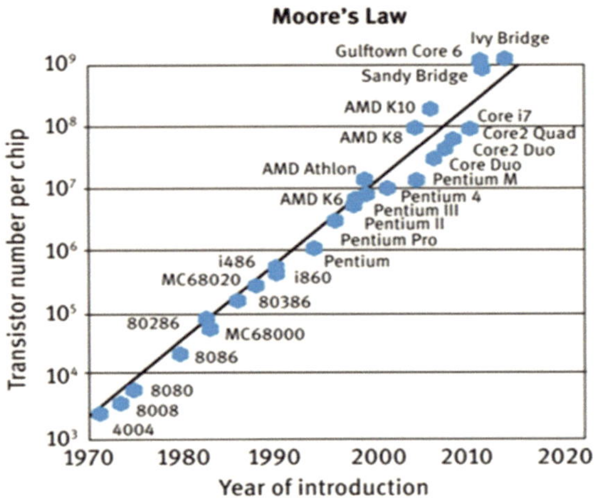

Figure 1. Graphical representation of Moore's law.

1.1. Need for low power design

Power dissipation is the main constraint when it comes to portability. Hence, it is necessary to take care of the system's total power consumption. Minimizing the overall power consumption in such devices is essential because it is advantageous to exploit the run time with least possible requirements on weight, battery life and size owed to batteries. Therefore, in portable devices, 'the low power design is the most decisive factor to think while designing system on chip. Normally, mobile users demand additional features and prolonged battery life at a lower cost. Almost 70% of users look for longer talk time and standby time as key feature for mobile phones. One of the top operator requirements in 4G is Power efficiency. Customers always look for smaller, trim and graceful mobile devices. This is the need of high levels of silicon integration in modern processes, but sophisticated processes have intrinsically higher power indulgence. So, design is very important in low power consumption devices.

1.2. Impact of power dissipation

Whenever there is power dissipation, it unvaryingly leads to an increase in chip temperature. This temperature rise affects devices when it is switched on and off. With device in OFF condition, power dissipation increases the number of intrinsic carriers n_i provided by the below relation:

$$n_i \; \alpha \; e^{-E_G/V_T} \tag{1}$$

From the above equation, it is very clear that when temperature increases, intrinsic carriers also increase. With temperature increase, the less affected ones are the majority carriers which are contributed by impurity atoms. As the temperature increases further, the leakage current that depends on the concentration of the minority carrier, increases which leads to further increase in temperature. Ultimately, the device might break down, if the dissipated heat is not removed properly. An ON device will not be affected much by the increase of

http://dx.doi.org/10.5772/intechopen.73729

minority carrier, but will be affected by the threshold voltage (V_T) and mobility (μ). These parameters decrease with increase in temperature and this leads to change in drain current (I_D). Hence the device performance might not meet the required specifications. Also, power dissipation is more critical in battery-powered applications as the greater power dissipated, the battery life will be less.

1.3. Reduction of temperature

Heat sinks are used to dissipate heat generated by power dissipation. The thermal resistance of heat sink is lower than that of the package. So heat sink draws the heat. To eliminate heat efficiently, the rate of heat transferred to the environment should be greater than heat generated. This heat transfer rate depends on thermal resistance θ, as provided by the below relation:

$$\Theta = l/\sigma_C A \tag{2}$$

where:

l is the length, A is the area and σc is the thermal conductivity of the heat sink.

From the above relation, it can be seen that large σc implies smaller θ. θ is also given by the relation

$$\Theta = \delta T/\delta P \tag{3}$$

Using this relation, we can see that for a given power dissipation, P_D

$$\Theta \leq \left(T_j - T_a \right)/P_D \tag{4}$$

where Tj is the junction temperature and Ta is the ambient temperature.

Heat sink materials are generally coated black to radiate more energy.

1.4. Low power design methodology

Historically, VLSI designers have used circuit speed as the performance metric. In fact, power considerations have been the ultimate design criteria in special portable applications. The main aim of these applications was maximum battery life time, with minimum power. Low power design is also required to reduce the power in high-end systems with huge integration density and thus improve the speed of operation.

To optimize power dissipation specifically with low power methodology in digital systems, the method should be applied all over the design from system to process level. It is very important to have knowledge about the power distribution. So the blocks or parts consuming fraction of power could be clearly optimized for saving power. Different design levels specifically of power reduction are shown in **Figure 2**.

1.4.1. Power reduction through process technology

Minimizing the supply voltage of a device is one of the best solutions to reduce power dissipation. The trade-off of this approach is that delay may increase significantly, when V_{DD}

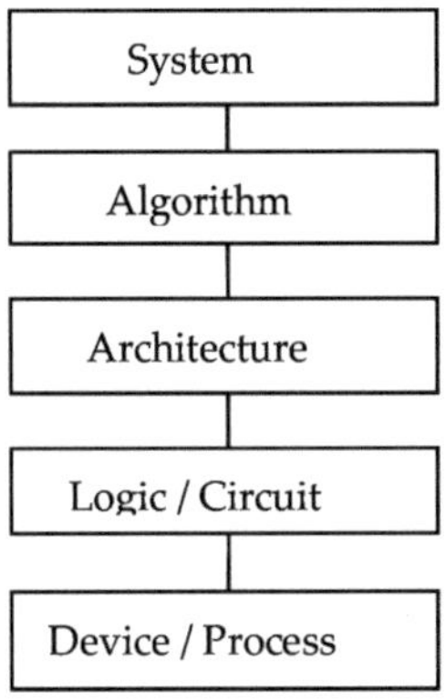

Figure 2. Power reduction design aspects.

approaches the threshold voltage. So devices must be properly scaled to overcome this problem. The advantages of scaling are:

- Improve the device characteristics
- Reduce the geometric and junction capacitances
- Enhanced interconnect technology
- High density of integration

1.4.2. Power reduction through circuit/logic design

- Use of more static than dynamic circuits
- Reduce switching activity by optimized algorithm
- Optimize clock and bus loading
- Smart circuit techniques which minimizes no of devices used in the circuit
- Custom design may improve the power
- Reduces VDD in non-critical paths and proper transistor sizing
- Use of multi-VT circuits
- Re-encoding of sequential circuits

1.4.3. Power reduction through architectural model

- Techniques for power management like shut down of unused blocks
- Architectures based on pipelining, parallelism etc.,
- Memory partitioning by enabling selective blocks
- Reduction in the numbers of global busses
- Instruction set minimization for easier decoding and execution

http://dx.doi.org/10.5772/intechopen.73729

1.4.4. Power reduction by algorithm level

- Minimizing the number of operation and hence reduce the number of hardware resources
- Data coding for reduce the switching activity.

1.4.5. Power reduction through system integration

- Utilize low system clocks
- Use high level of integration

1.5. Power modelling

Numerous power components and their outcome must be identified to reduce power consumption of certain circuit. Out of two power dissipation types, the maximum power dissipation relates to peak instantaneous current and the second type is average power dissipation. Due to power line resistance, peak current affects the noise in supply voltage. This causes heating of device and hence results in performance degradation. With a view on battery life time, this average power dissipation becomes more important. The three important power dissipation components are [1]

- Static power due to leakage current I_{Leak} and other static component I_{St} due to the value of the input voltage
- Dynamic power caused by the total output capacitance C_L and short circuit current I_{SC}, during the switching transient
- Short circuit power dissipation

Thus the total power dissipation P_T is

$$PT = PS + PD + PSC$$

1.5.1. Static power dissipation

Static power dissipation is the power consumed during the standby mode of a design. CMOS gates typically have some amount of sub-threshold leakage current even when gates are not turned on. The drain to source leakage current is the main component of static power consumption. The leakage power is a very small part of the overall power consumption. In a typical chip 10% of the power consumed is leakage and 90% is dynamic power. So, clearly the major concern is dynamic power dissipation. **Figure 3** shows static power calculation model.

$$\text{Instantaneous power } P(t) = i_{DD}(t)V_{DD} \tag{5}$$

$$\text{Energy } E = \int_{0}^{T} p(t)dt \tag{6}$$

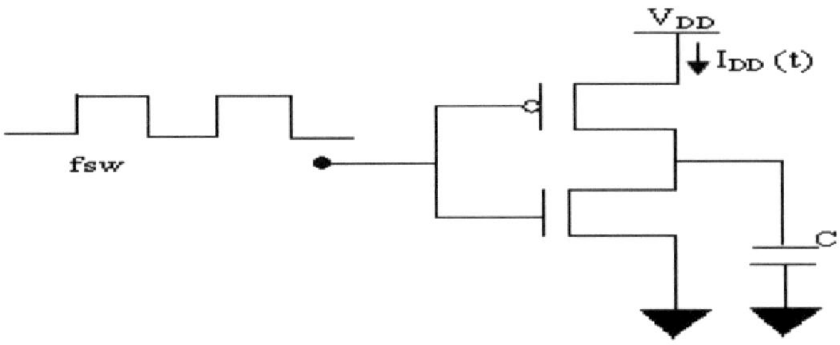

Figure 3. Static power calculation model.

$$E = \int_{0}^{T} iDD(t)VDD\,dt \tag{7}$$

$$\text{Static power } P_S = E/T \tag{8}$$

$$P_S = 1/T \int_{0}^{T} iDD(t)VDD\,dt \tag{9}$$

1.5.2. Dynamic power dissipation

A dynamic power vector describes an event in which power is dissipated due to a signal switching at the cell input during charging and discharging of load capacitance. Dynamic power is further divided into switching power and internal power.

- Switching power

Switching power is dissipated when the load capacitance at the output of the cell is being charged or discharged. The load capacitance is composed of interconnect capacitance and gate capacitances. Switching activity of cells depend on the quantity of switching power. On the cell output, if there are huge logic transitions, then switching power surges.

- Internal power

Within a cell, internal power is specifically consumed for charging and discharging cell capacitances. When logical transitions occur, Pmos and Nmos transistors are ON at the same time for a short period. This causes a connection between Vdd and ground rails.

The power dissipation can be estimated by the load capacitance C_L. This power loss is due to the charging and discharging of load capacitance C_L [1]. The average dynamic power P_D is required to charge and discharge a capacitance C_L at a switching frequency fsw and equivalent dynamic power calculation model is shown in **Figure 4**.

$$P_D = f_{sw} \int_{0}^{T} io(t)Vo(t)dt \tag{10}$$

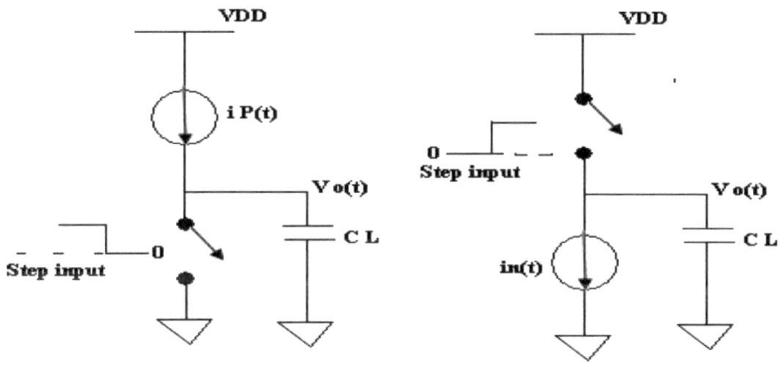

Figure 4. Equivalent circuits for dynamic power calculation.

During charging cycle

$$i_p = C_L.\frac{dVo}{dt} \tag{11}$$

During the discharge cycle

$$i_n = -C_L.\frac{dVo(t)}{dt} \tag{12}$$

$$\mathrm{P_D} = \mathrm{fsw}\left[\int_0^{VDD} C_L V_o dVo - \int_{VDD}^{0} C_L V_o dVo\right] \tag{13}$$

$$\mathrm{P_D} = \mathrm{fsw}\left[\mathrm{C_L}\left[\left[\frac{Vo^2}{2}\right]_0^{VDD} - \left[\frac{Vo^2}{2}\right]_{VDD}^{0}\right]\right] \tag{14}$$

$$\mathrm{P_D} = \mathrm{fsw}\left[\mathrm{C_L}\left[\frac{V_{DD}^2}{2} + \frac{V_{DD}^2}{2}\right]\right] \tag{15}$$

$$\mathrm{P_D} = \mathrm{fsw}\ \mathrm{C_L}\ \mathrm{V_{DD}}^2 \tag{16}$$

Assuming a logic gate goes through one complete charge/discharge cycle for every clock cycle, suppose the system clock frequency is f.

Let fsw = Ef, where E is the energy transition activity factor.

Most gates do not switch every clock cycle,

$$\mathrm{P_D} = \mathrm{E\ C_L.V_{DD}}^2.\mathrm{f} \tag{17}$$

A clock has E = 1 because it rises and fall every cycle, but most data have a maximum energy transition activity factor E = 0.5 because they transit only once every cycle.

The dynamic component of power consumption arises when the capacitive load C_L of a CMOS circuit is charged through PMOS transitions to make a voltage transition from 0 to 1, half of which is stored in the output capacitor and half is dissipated in the PMOS device [2]. No

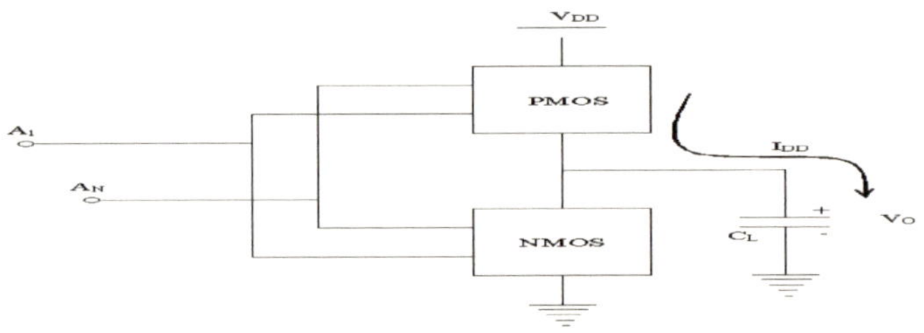

Figure 5. Energy per transition.

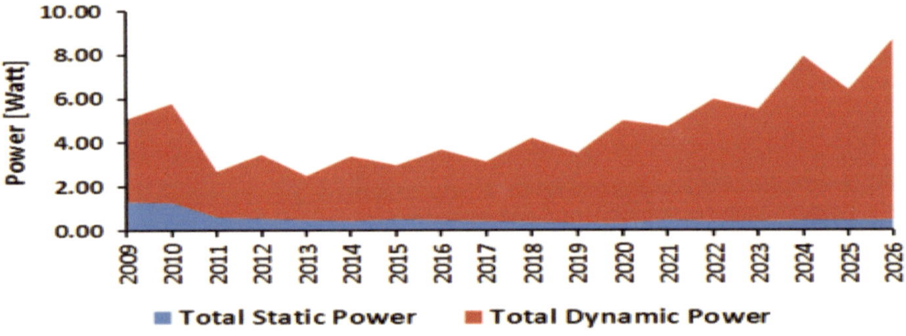

Figure 6. Power analysis chart, IRTS-2011.

charge is drawn from the V_{DD} during the 1 to 0 transition at the output. But the energy stored in the capacitor is dissipated in the pull-down NMOS device shown in **Figure 5**. The main cause of energy dissipation in CMOS circuits is due to charging and discharging of the node capacitances. The power analysis chart is also shown in **Figure 6**.

1.6. Short circuit power dissipation

Short circuit current occurs during signal transitions when both the NMOS and PMOS are ON and there is a direct path between Vdd and GND. Also called crowbar current, the total power dissipation is more than 20% of total power. As clock frequency increases, transitions increases and consequently short circuit power dissipation increases. It can be reduced by

- faster input and slower output
- $Vdd <= Vtn + |Vtp|$

So both NMOS and PMOS are not ON at the same time.

The short-circuit power dissipation is given by

$$P_D = I_{mean}{}^{*} V_{DD} \quad (18)$$

For the input waveform shown in **Figure 8**, which depicts the short circuit in an unloaded inverter,

$$I_{mean} = 2x\frac{2}{T}\int_{t_1}^{t_2}\frac{\beta}{2}(V_{in}(t) - V_t)^2 dt \tag{19}$$

assuming that $V_{tn} = -V_{tp}$ and $\beta_n = \beta_p$ and that the behaviour is symmetrical around t2.

$$I_{mean} = 2x\frac{2}{T}\int_{t_1}^{t_2}\frac{\beta}{2}(V_{in}(t) - V_t)^2 dt \tag{20}$$

With

$$V_{in}(t) = \frac{V_{DD}}{t_r}t \tag{21}$$

$$t_1 = \frac{V_t}{V_{DD}}t_r \tag{22}$$

$$t_2 = \frac{t_r}{2} \tag{23}$$

$$t_r = t_f = (t_{rf}) \tag{24}$$

Assuming an inverter without load,

$$P_{SC} = \frac{\beta}{2}(V_{DD} - 2V_t)^3\frac{t_{rf}}{t_p} \tag{25}$$

where tp is the period of the waveform.

The equation suggests that, depending on the input rise and fall times and β, the short circuit current varies. For load inverters, on nodes, slow rise times significantly reduces (20%) SC power dissipation. If power dissipation is a concern, then it is good if all the edges are kept fast. Further increase in load capacitance significantly reduces the short circuit dissipation by reduced capacitive dissipation P_D.

1.7. Transition activity

The internal power and the capacitive load power are the two key components for dynamic power dissipation in a complex design, like the internal node. The power in an internal node is determined by the amount of the power dissipated by the internal capacitive nodes [3]. Sometimes, internal node short circuit power is also included in the node to calculate the dynamic power at the internal node. So the dynamic power cannot be calculated by the simple equation $C_L V_{DD}^2 f$ because MOS devices might not switch when the clock is switching. The transition activity determines how often this transition occurs on a power. Considering capacitive node for N periods of time $0 \rightarrow 1$ and $1 \rightarrow 0$ transitions will occur. The transition activity E determines how many low to high and high to low transitions occur at the output [4]. In other words, the activity E represents the probability that a transition $0 \rightarrow 1$ will occur during the period $T = 1/f$. The average dynamic power of a complex design due to the output load capacitance is given by

$$P_D = E\,C_L\,V_{DD}^2 f \quad (26)$$

The internal power dissipation, due to internal nodes, the internal dynamic power of a cell is given by

$$P_{int-dyn} = \sum_{i=1}^{\propto} E_i C_i V_i\, V_{DD}\, f \quad (27)$$

Due to charging and discharging the data changed from 1 to 0 or from 0 to 1 vice versa between adjacent bus wires or on the same bus wire. This is classified into two types:

- Self-transition
- Coupling transition

1.7.1. Self-transition

A Self-transition (ST) is defined as a transition from $0 \rightarrow 1$ or $1 \rightarrow 0$ on bus with reference to the previous data on it [5]. Energy transition analysis is shown in **Table 1**.

1.7.2. Coupling transition

A coupling transition (CT) is defined as a transition from $0 \rightarrow 1$ or $1 \rightarrow 0$, between two adjacent bus wires [5]. The corresponding energy transition analysis is shown in **Table 2**.

1.8. Design parameter

The low power design work mainly focuses on estimating the dynamic power dissipation. In the past, the major concern of the designer was about area, speed and cost. The secondary importance was provided for power considerations. In recent years, power has become as the primary

Transition of bits	State	Energy stored initially	Energy stored finally	Energy dissipated	Energy consumed
0 → 1	charge	0	$E_S/2$	$E_S/2$	E_S
1 → 0	discharge	$E_S/2$	0	$E_S/2$	0

Table 1. Energy transition analysis for self-capacitance, Yan Zhang et al. 2002.

Transition of bits	State	Energy stored initially	Energy stored finally	Energy dissipated	Energy consumed
00 → 00	-	0	0	0	0
00 → 01	charge	0	$E_C/2$	$E_C/2$	E_C
00 → 10	charge	0	$E_C/2$	$E_C/2$	E_C
00 → 11	-	0	0	0	0
01 → 00	discharge	$E_C/2$	0	$E_C/2$	0
01 → 01	-	0	0	0	0
01 → 10	Toggle	$E_C/2$	$E_C/2$	$2E_C$	$2E_C$
01 → 11	discharge	$E_C/2$	0	$E_C/2$	0
10 → 00	discharge	$E_C/2$	0	$E_C/2$	0
10 → 01	Toggle	$E_C/2$	$E_C/2$	$2E_C$	$2E_C$
10 → 10	-	0	0	0	0
10 → 11	discharge	$E_C/2$	0	$E_C/2$	0
11 → 00	-	0	0	0	0
11 → 01	Charge	0	$E_C/2$	$E_C/2$	E_C
11 → 10	Charge	0	$E_C/2$	$E_C/2$	E_C
11 → 11	-	0	0	0	0

Table 2. Energy transition analysis for coupling capacitance, Yan Zhang et al. 2002.

design consideration. Several factors contribute to this trend like the growth of personal computing devices such as portable desktops, audio and video-based multimedia products and wireless communication systems which demand high-speed computation and complex functionality with low power consumption [6]. So there is a strong requirement for power consumption reduction so as to reduce packaging and cooling cost and improve product reliability. When the target is a low power application, a power analyser/estimator ranks the various design aspects, thus helps in selecting the one that is potentially more effective from the power standpoint.

1.8.1. Two-dimensional design flow

A top-down two-dimensional ordinary VLSI design approach is illustrated in **Figure 7**. The figure summarizes the flow of steps that are required to follow from a system-level specification to the physical design. The approach is aimed to estimate the design parameters such as the performance optimization and area minimization, as shown in **Figures 8–10**.

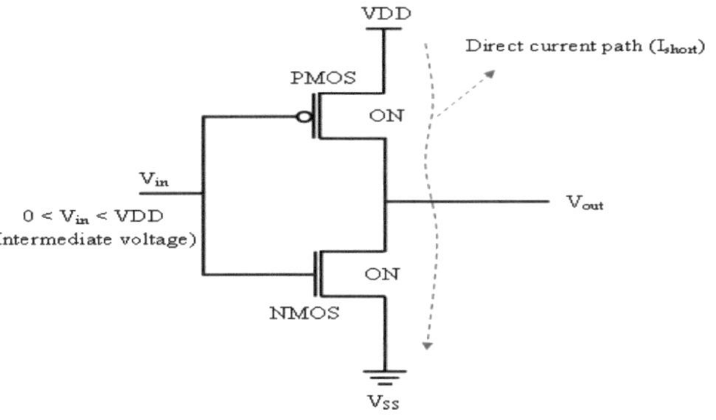

Figure 7. Short circuit power calculation model.

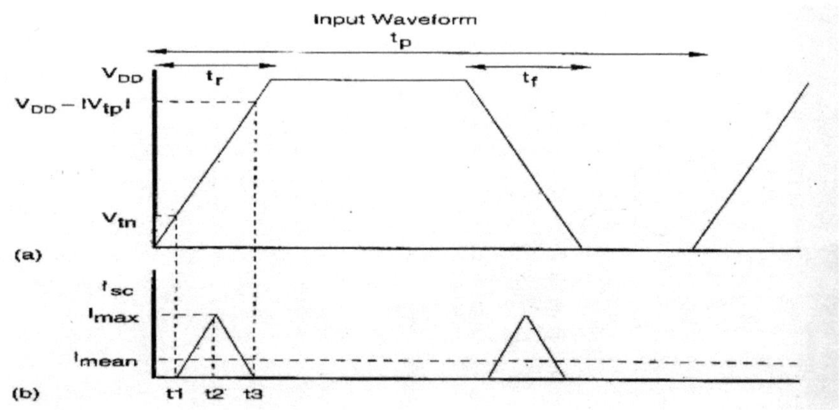

Figure 8. Short circuit behaviour of CMOS inverter without load.

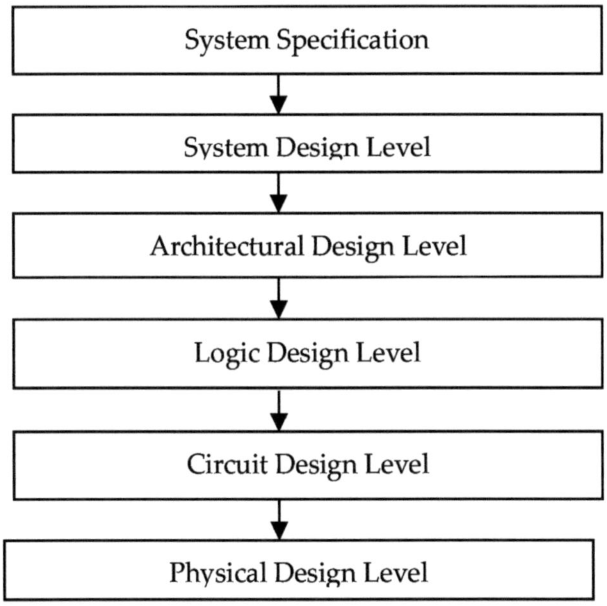

Figure 9. Two-dimensional (2D) VLSI design flow.

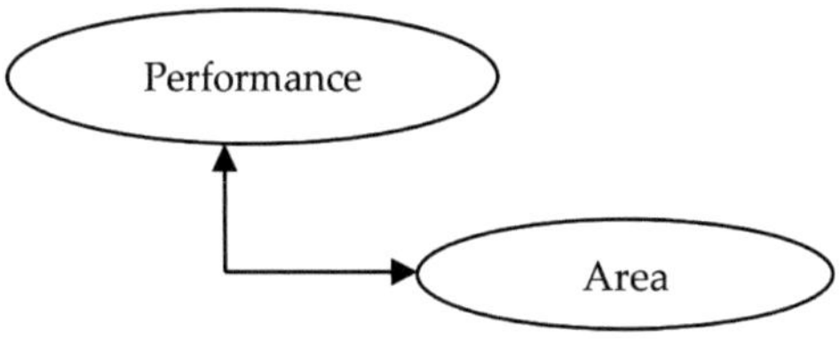

Figure 10. Two-dimensional (2D) design parameter.

1.8.2. Three-dimensional design flow

A three-dimensional top-down VLSI design approach is illustrated in **Figure 11**. The figure summarizes the flow of steps that are required to follow from a system-level specification to the physical design. The approach is aimed to estimate the design parameters at performance optimization, area minimization and power optimization shown in **Figure 12**. In each of the design levels, there are two important power factors, namely, power optimization and power estimation. Power optimization is the process of obtaining the best design knowing the design constraints and without violating design specifications. Power estimation is determined as the process of computing power and energy dissipated with a definite percentage of precision and at different stages of design process. This technique also estimates the outcome of several optimization and design alterations on power at different levels of abstraction, as shown in **Figure 12**. Design attains power optimization first and then does power estimation. But for certain design, there is no specific design procedure. Each design might include a lot of low power techniques and thus significantly reduce power dissipation. But certain combination of low power designs can provide better result than certain other combination techniques. Usually

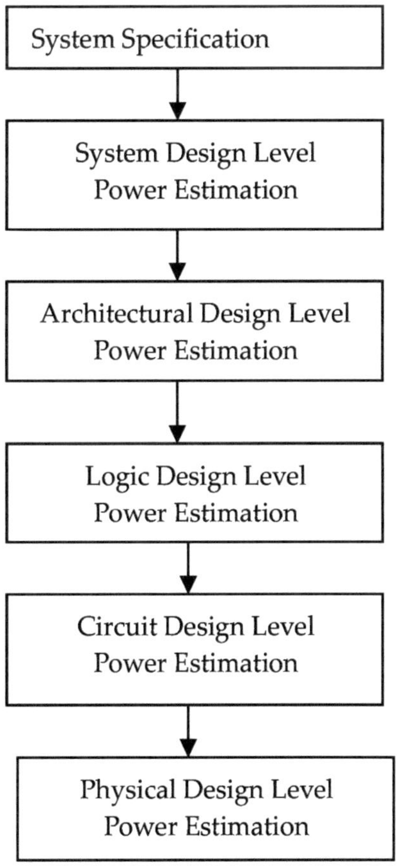

Figure 11. Three-dimensional (3D) VLSI design flow.

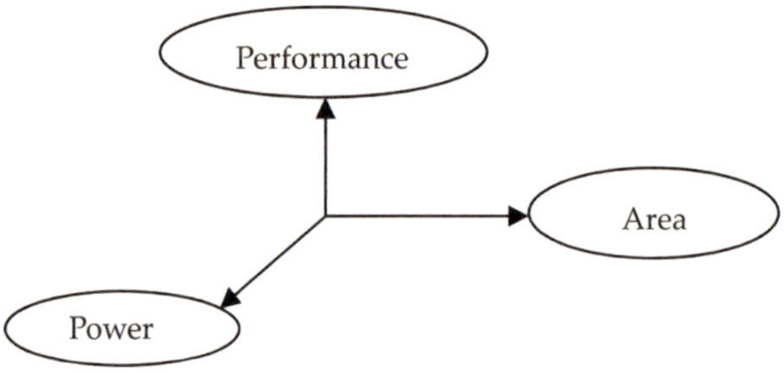

Figure 12. Three-dimensional (3D) design parameter.

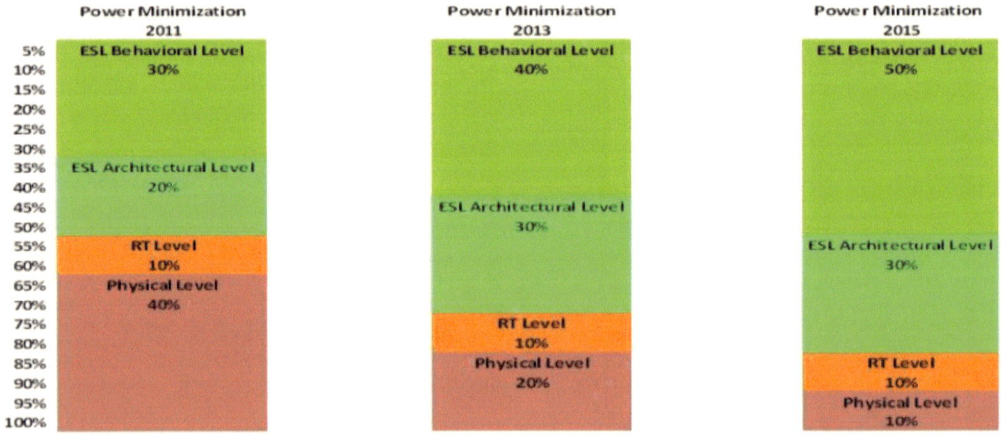

Figure 13. Relationship between different abstraction level and power estimation techniques.

power will be consumed due to transition activities as the capacitors gets charged and discharged. So for higher level systems, power dissipation is preserved by shutdown of system portions when not required and thus the transition activities are reduced (**Figure 13**).

1.9. Power estimation tool

Recently, complexity levels of device size and programmable devices have grown to amazing complexity levels. Years ago, an average design had nearly twelve thousand gates. Presently, there are hundreds of thousands and sometimes multimillion gates. So when size of design increases, power consumption also increases. In the meantime, there is huge demand for battery-powered systems, specifically, handheld devices which are constantly sensitive and smaller to power usage. So it is clearly understood that in programmable logic devices design power consumption cannot be ignored. This chapter deals more on power calculations using Macros and is experimented using power tools. Prior to the power tools, other tools have been used to provide the necessary input to the power tools. More importance is provided to the tools specifically involved in low power estimation, which has been classified as power tools and non-power tools.

1.9.1. Non-power tool

Non-power tools include simulation tools, synthesis tools, layout tools, extraction tools and waveform viewers.

1.9.2. Power tool

Varieties of power analysis tools are available to estimate the power of a design. Among them are Xilinx, Tanner, Microwind, etc. These EDA power tools are very familiar and user-friendly. The power products are tools that comprise a complete methodology for low power design. Xilinx power tool XPower offers power analysis and optimization throughout the design cycle (from RTL to the gate level). Tanner and Microwind are used for transistor-level analysis. Analysing power early in the design cycle can significantly affect design quality. Design modifications done at RTL level can get good results. Power tools used to calculate power quickly as well as do measurements accurately. The following tools are used to calculate the power at these levels.

a. Tanner EDA, Microwind: Transistor level

b. RTL Power Estimator: RTL level

c. Power Compiler: Gate level.

Power analysis and estimation is available throughout the design process, as shown in **Figure 14**.

- XPower analysis tool

Activity rates are the basis of Xilinx Power tool. They are defined by the rate at which a logic element or net capacitance switches. Activity rates for dynamic calculations are expressed in

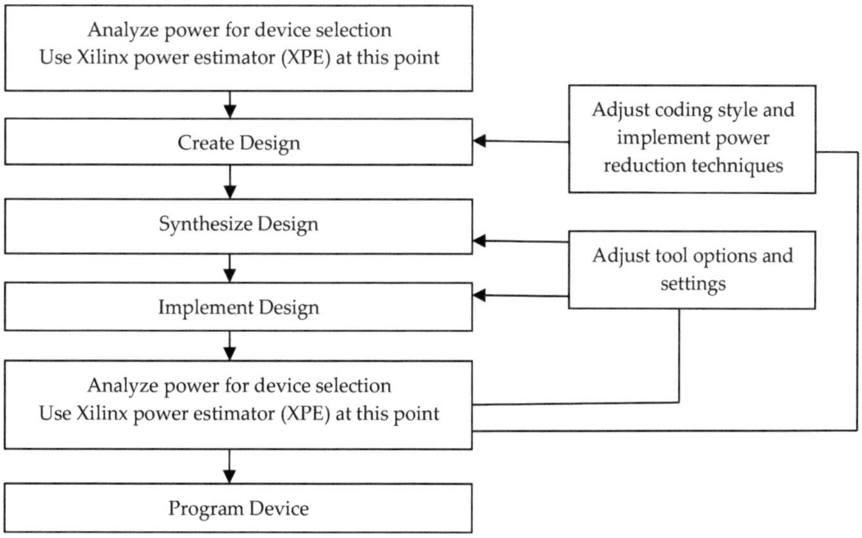

Figure 14. Power analysis flow chart.

frequency. The activity rate might be relative to clock and hence net or logic element might switch at any fraction of the clock frequency. Thus the main use of activity rate is in the recalculation of power and could be easily achieved by varying system clock frequency. So simulation data could be used, and this saves time. Also Xilinx Power supports several numbers of input clocks. Expressed in percentage scale, 100% activity rate means that standard signal state changes once every clock cycle. Switching rate will be the activity rate if net and logic are not clock sync (**Figures 15–22**).

- Microwind

This software tool is dedicated to microelectronics and nanotechnology. The microwind software allows the designer to simulate and design an integrated circuit at physical description level. It provides innovative EDA solutions to the analog, digital and mixed-signal IC market. With MOS characteristic viewer, mix signal simulator, in-built layout editing tools, it is easier to complete design process. Microwind unifies netlist extraction, pattern-based simulator, layout compilation, SPICE extraction of schematic, Verilog extractor, schematic entry on layout

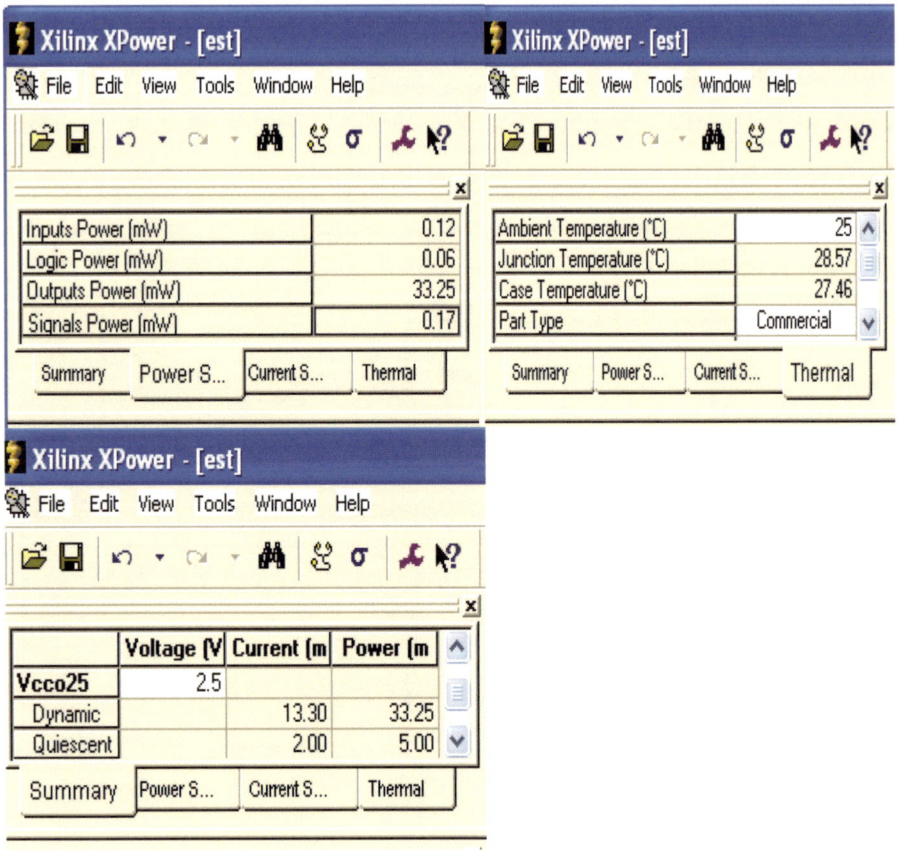

Figure 15. Power output calculation using XPower.

http://dx.doi.org/10.5772/intechopen.73729

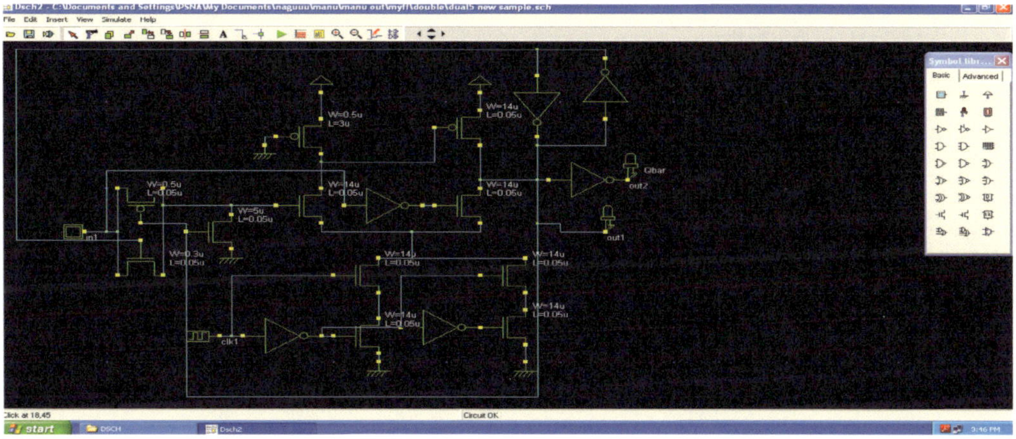

Figure 16. Simulation of digital CMOS circuits.

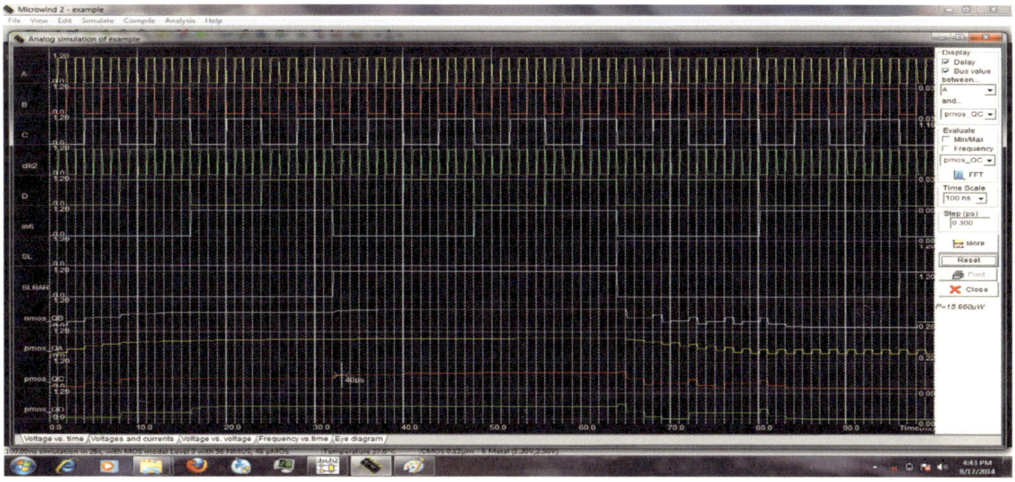

Figure 17. Power calculation of digital CMOS circuits.

mix-signal circuit simulation, sign-off correlation, BSIM4 tutorial on MOS devices, cross-sectional and 3D viewer to deliver matchless architecture productivity and performance.

- Tanner EDA Tool

Tanner tool is a suite of tools to perform spice analysis for analog integrated circuits. Following are the Tanner tool engine machines:

1. Schematic Edit (S-EDIT)
2. Simulation Edit (T-EDIT)
3. Waveforms Edit (W-EDIT)
4. Layout Edit (L-EDIT)

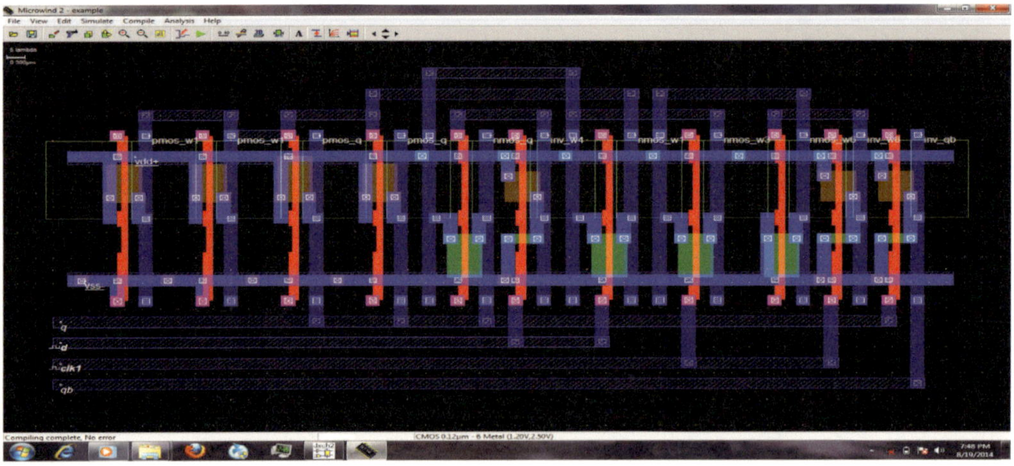

Figure 18. Layout of digital CMOS circuits.

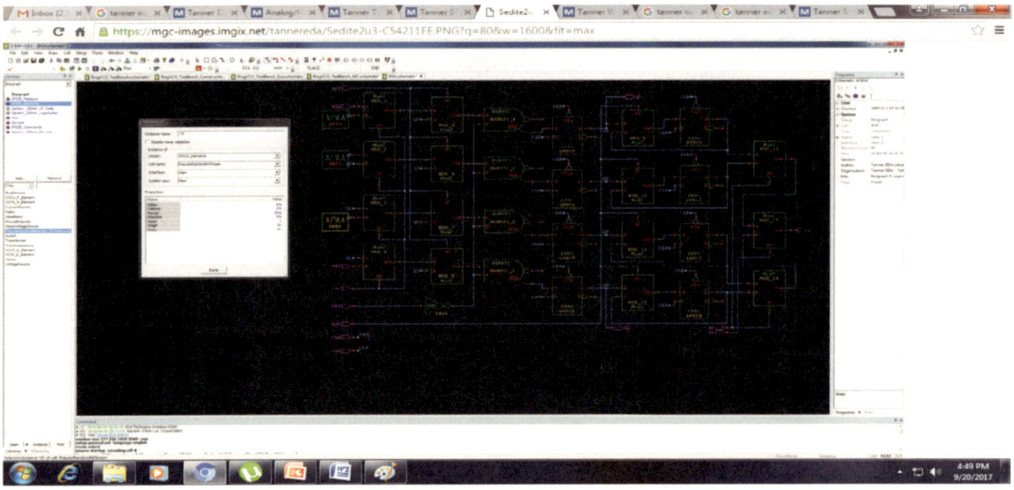

Figure 19. Tanner S-Edit schematic capture.

The Tanner engine tools are used to design and simulate new ideas in analog-integrated circuits; this saves time and cost of chip fabrication.

1.10. Conclusion

In CMOS circuits, most of the power dissipates through dynamic power dissipation than static power dissipation. In CMOS circuits, static power dissipation is in the range of nano watts. The most significant source of dynamic power dissipation is caused by transition activities of the circuits. A higher operating frequency leads to more transition activities in the circuits and results in increased power dissipation. Using proper encoding techniques may reduce switching activity in the circuit. This will reduce the overall transition activity. Hence, the dynamic power dissipation can be reduced in VLSI circuits effectively.

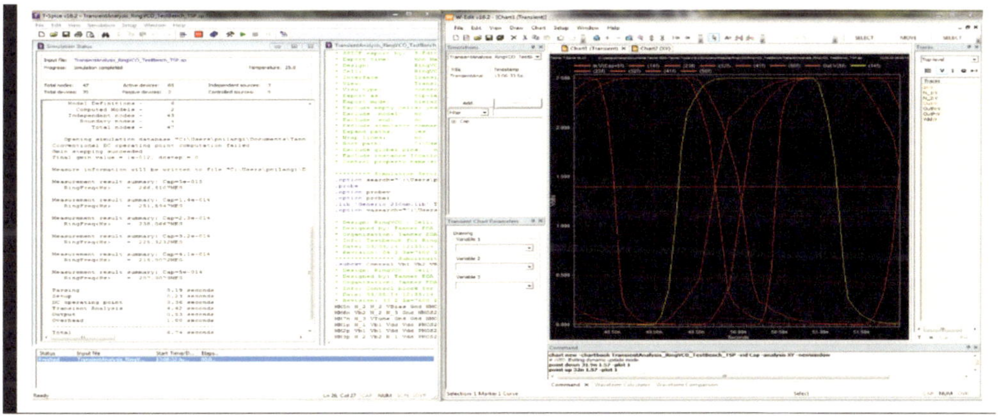

Figure 20. T-spice simulation.

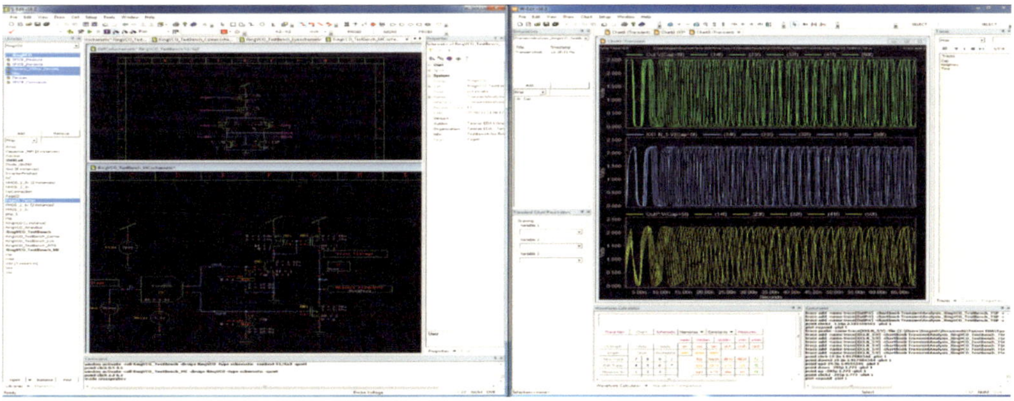

Figure 21. Tanner waveform viewer.

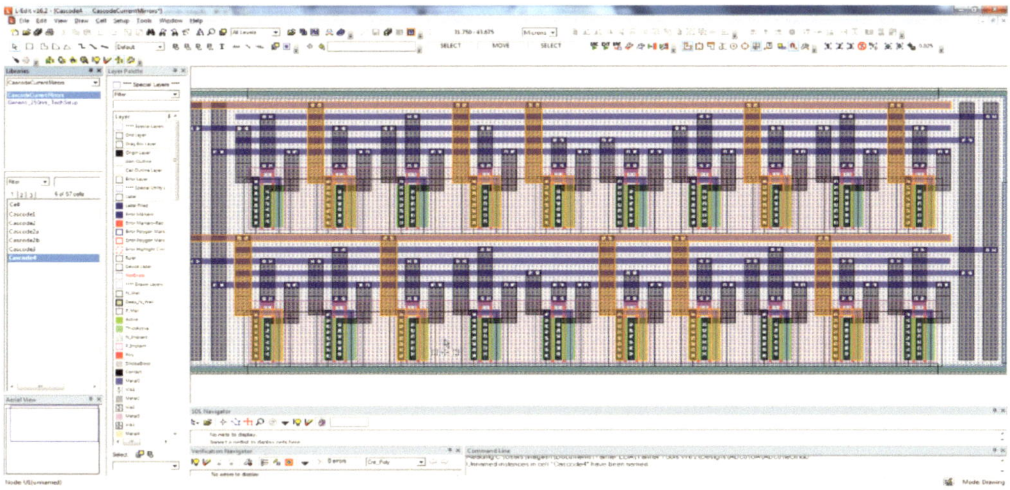

Figure 22. L-Edit IC layout.

Author details

Vithyalakshmi Natarajan[1]*, Ashok Kumar Nagarajan[1], Nagarajan Pandian[1] and Vinoth Gopi Savithri[2]

*Address all correspondence to: vidhyavinoth@gmail.com

1 Sree Vidyanikethan Engineering College, Chittoor, Andhra Pradesh, India

2 QuEST, Global, Technopark, Trivandrum, Kerala, India

References

[1] Weste N, Eshraghian K. Principle of CMOS VLSI Design: A System Perspective. 2nd ed. New York: Addison–Wesley; 1993

[2] Stan MR, Burleson WP. Bus-invert coding for low power I/O. IEEE Youngsoo Shin, Soo-Ik Chae & Kiyoung Choi 2001. Partial Bus-Invert Coding for Power Optimization of Application-Specific Systems. IEEE Transactions On Very Large Scale Integration Systems. 1995, vol. 9, no. 2, pp. 377-383. Trans. On VLSI, vol. 3, pp. 49-58

[3] Young Chul Kim, Young Jin Lee. Power effective bus encoding scheme with no crosstalk and minimized bus Transition. International Journal of Control and Automation. 2013;**6**(4):1-10

[4] Verma SK, Kaushik BK. Encoding schemes for reduction of power dissipation, crosstalk and delay in VLSI interconnects: A Review. International Journal of Recent Trends in Engineering and Technology. 2010;**3**(4):74-76

[5] Sainarayanan KS, Ravindra JVR, Kiran T Nath, Srinivas MB. Coding for minimizing energy in VLSI interconnects. The 18th International Conference on Microelectronics. 2006 pp. 166-169

[6] Abhijeet Dhanotiya, Vishal Sharma. Power reduction in digital VLSI circuits. International Journal of Research in IT, Management and Engineering. 2014;**4**(6):13-23

Chapter 4

High-purity Refractory Metals for Thin Film Metallization of VLSI

Vadim Glebovsky

Additional information is available at the end of the chapter

http://dx.doi.org/10.5772/intechopen.69126

Abstract

It is shown that cast targets of highly pure refractory metals like W, Mo, Ti, Ta, Co, etc. and their compounds can be produced by means of a set of vacuum-metallurgical techniques—by vacuum high-frequency levitation, EB floating zone melting, EB melting, and electric arc vacuum melting as well as chemical purifying by ion exchange and halides. The cast refractory metal targets are extremely pure and chemically homogeneous. For magnetron sputtering and laser ablation, the cast silicide targets are also produced. The study reveals the possibilities and conditions of depositing the silicides and titanium-tungsten barrier layers by both the laser evaporation and magnetron sputtering. The physical and structural parameters as well as a trace impurity composition of sputtered metals and deposited thin films are studied by grazing-beam incidence X-ray diffraction, Auger electron spectroscopy, Rutherford backscattering of helium ions, mass spectrometry with inductively coupled plasma, etc.

Keywords: EB vacuum melting, target, magnetron sputtering, laser ablation, thin films, resistivity, refractory metals

1. Introduction

The application of thin films of pure Al and Al doped with Si (1 at.% Si) seems to be useful for the production of thin films for integrated circuits with dimensions of about 2 μm. A further reduction of dimensions to less than 1 μm and decreasing the *p-n* transition to 0.3 μm increase demands for parameters/operation conditions and necessitate a search for more reliable constructive and technological designs of the metallization including the contact units. Hence, ultra-purity is a term increasingly being applied to refractory metals [1–3]. Not only applications in the electronic industry but also their use as superconducting materials or applications in the nuclear industry call for refractory metals and their alloys of utmost purity. The most intensive impact on ultra-purity of refractory metals is presently exerted by the electronic industry for VLSI applications [4–6]. In a word, VLSI means miniaturization. The goal of VLSI development is to find ways to pack as many tiny electronic components as possible into the smallest possible space. Often, smaller circuits also operate faster and, ultimately, drastically cut the overall cost of computation, opening up new possibilities for applications. Continued evolution of ever smaller devices has aroused a renewed interest in the development of new metallization processes for low resistivity gates, interconnections, and ohmic contacts. Al, W, and Mo are notable among the metals proposed for gate and interconnection metallization. The use of Al, however, requires all postgate processing of devices to be limited to very low temperatures, preferably below 500°C. The use of refractory metals, such as W and Mo, requires a complete passivation of these metals against oxidizing environments, a deposition by means that will not lead to unwanted traps in the gate oxide, and reliable etching of metals for pattern generation. Uncertainties associated with the stability of these metal films have led to a search for alternatives. As a rule, the different components deposited onto the Si substrate by different sputtering/evaporation methods must be ultrapure, especially with respect to trace contaminants listed in **Table 1** where the impurity compositions are represented for refractory metals studied in this chapter. The wide variety of applications of ultrahigh purity refractory metals as sputter targets in microelectronics is really impressive. A great number of companies compete to offer ultrapure refractory metal materials for sputter targets. Materials presently in the microelectronics use are W/Ti, Mo, $MoSi_2$, WSi_2, $TaSi_2$, $TiSi_2$, etc. As a rule, different components must be ultrapure, especially with respect to "mobile ions" (Li^+, Na^+, K^+, Ca^+, and Mg^+). With increasing miniaturization, an intrinsic radioactivity also exerts harmful effects; therefore, various components must be "free" of U and Th contamination. Nonmetals (O, N, H, and C) are detrimental too, as are impurities of Fe, Co, and Ni, for certain applications. However, information on analytical methods used for a chemical characterization of such materials as well as on metallurgical techniques used for a preparation of materials is extremely scarce or sometimes also misleading. It is known that a metal gate cannot withstand to the oxidizing annealing ambient and a source-drain formation by ion implantation is difficult because of the channeling of doping ions through the gate metal during ion implantation. In the process developed for MOS VLSI fabrication, W is used as the gate metal because a degradation of SiO_2 by annealing a Me/SiO_2/Si structure at about 1000°C can be minimized. An oxidation of W is prevented by a moist hydrogen atmosphere during annealing. Si is also oxidized in the similar ambient. The most effective solution for VLSI is by applying barrier layers, such as Ti/W thin layers, nitrides, or silicides of refractory metals. The use of barrier layers requires a rational technique for their deposition as well as the materials of an optimal chemical composition. Even using Ti/W thin films as both barrier and conducting layers, the most suitable and advanced deposition technique is the magnetron sputtering, which is widely employed in commercial microelectronics. Thin films of Ti/W quasi-alloys as diffusion barriers for metal contacts on Si are used very intensively in the last decades. The W in this composition is supposed to serve as an interlayer diffusion barrier and the Ti as both a deoxidizer and a stopper of the grain boundary diffusion. The well-known channeling is stopped by forming a thin

Impurity	Mo	Ti	Ta	W	Nb	V	Zr	Hf	Co	Ni
C	0.1	30.0	30.0	4.0	40.0	20/0	50.0	50.0	10.0	0.5
O	0.5	400.0	5.0	1.0	5.0	200.0	100.0	10.0	20.0	–
H	0.2	–	2.0	0.1	1.0	–	–	–	1.0	–
N	0.05	5.0	5.0	5.0	5.0	5.0	5.0	5.0	7.0	–
Fe	0.5	10.0	0.5	1.0	0.5	8.0	300.0	3	50.0	0.1
Al	0.03	5.0	2.0	0.04	2.0	20.0	10.0	0.05	10.0	0.1
Cu	0.6	0.4	0.6	0.3	0.5	0.08	300.0	4	10.0	0.1
Ni	0.2	10.0	3.0	1.0	5.0	0.3	100.0	0.3	370.0	Matrix
Ti	0.2	Matrix	3.0	–	3.0	0.3	–	2.0	10.0	0.2
Si	0.3	0.6	1.0	0.04	5.0	2.0	1.0	0.5	15.0	1.0
S	0.3	0.2	0.2	0.1	5.0	2.0	–	–	10.0	–
P	1.0	1.0	1.0	0.5	1.0	0.5	–	–	5.0	–
Nb	2.0	2.0	30.0	1.0	Matrix	0.2	–	–	4.0	1.0
W	20.0	100.0	2.0	Matrix	30.0	0.3	–	0.3	10.0	1.0
Co	–	–	0.2	–	2.0	0.3	–	–	Matrix	0.1
Mo	Matrix	–	0.2	0.5	4.0	–	0.3	0.5	10.0	1.0
Ta	2.0	–	Matrix	1.0	50.0	30.0	–	–	5.0	1.0
Cr	–	0.3	0.3	–	0.5	0.3	5.0	1.0	10.0	0.1
Cd	1.0	0.7	0.3	0.1	0.3	0.2	–	–	0.1	0.2
Mg	–	0.5	0.2	0.1	1.0	2.0	10.0	0.1	10.0	0.1
K	0.4	8.0	1.0	0.02	1.0	1.0	1.0	–	1.0	0.3
Na	0.2	–	–	0.01	–	–	10.0	–	1.0	–
Li	–	–	–	–	–	–	10.0	–	1.0	–
Ca	1.0	1.0	0.3	1.0	1.0	1.0	1.0	1.0	10.0	0.1
Sb	1.0	0.7	0.3	1.0	0.3	0.5	1.0	–	1.0	–
Mn	0.2	0.3	0.3	0.03	0.3	0.1	1.0	0.1	10.0	0.1
Zr	0.3	0.3	0.3	–	0.3	0.3	Matrix	5.0	10.0	–
As	0.2	0.5	0.3	0.3	0.3	0.2	0.3	–	1.0	0.1
Pb	2.0	4.0	1.0	1.0	1.0	1.0	1.0	1.0	10.0	–
Zn	0.7	0.4	0.3	0.3	0.3	0.3	0.3	0.1	1.0	–
Sn	1.0	–	0.3	1.0	0.3	–	0.3	1.0	1.0	–
Bi	–	–	0.3	1.0	0.3	–	–	–	10.0	–

Impurity	Mo	Ti	Ta	W	Nb	V	Zr	Hf	Co	Ni
V	0.5	1.0	1.0	0.4	1.0	Matrix	1.0	1.0	10.0	–
U	0.005	0.006	–	0.005	–	–	–	–	0.005	0.05
Th	0.005	0.006	–	0.005	–	–	–	–	0.005	0.05

Table 1. Trace impurities in refractory metals participating in these studies.

layer of BPSG or WO_x on the W [4]. Because barrier layers in integrated circuits simultaneously serve as conducting layers, it is very important to have all possible information on their physical characteristics. However, electric characteristics of Ti/W contacts to Si, which have naturally to be dependent on the film stoichiometry, structure, and purity of Ti/W thin films, are not studied in a whole concentration range in the Ti/W system. The only experimental data available are results for Ti/W alloys with a mean content of 10–30 at.% W because these quasi-alloys are supposed to be optimal ones for microelectronics. Other problems aroused are connected with the preparation of sputter targets of a known chemical composition and required purity. A preparation of targets by the powder metallurgy (PM) techniques when compacted targets are manufactured using procedures such as powder mixing, hot pressing, hot rolling, and annealing is fraught with the contamination by a considerable amount of impurities, especially gas-forming ones. Thus, it can be said that the impurity content in targets prepared by PM techniques is predetermined by the preparation technique itself. As a result, there exists a possibility of a strong instability of the sputtering process due to an intensive gas release from the compacted powder target and even unpredictable fracture of sputtering components because of the explosive-like gas release at higher temperatures. It seems that a solution of this problem is in the combination of high-pressure techniques and in situ high-temperature annealing of PM compacted targets in a high vacuum [7]. Such a complex procedure allows one to produce PM compacted targets with a density which is nearly equal to the tabulated one of the material. An alternative to PM techniques is the use of cast targets produced from high-purity cast refractory metals without contaminating procedures like open-air hot pressing or hot rolling [8]. When it is necessary to prepare thin films of alloys, targets composed of cast metal blocks or co-sputtering of several cast metal targets can be used. It is known that TiW alloys cannot be produced by conventional melting procedures because of the great difference of their melting temperatures of both components. The sputtering or co-sputtering of cast metal targets seems to be the most promising ones owing to their flexibility and reliability, i.e., it is possible to obtain any necessary chemical composition of films due to the wide choice of compositions of "mosaic" targets, as well as deposition rates of each metal component during co-sputtering. One of the sections in the chapter will be discussing the experimental results of studies of the deposition of Ti/W thin films with different Ti/W ratios, made by magnetron sputtering of two cast metal targets, as well as of the dependence of the electrical resistivity on the Ti/W ratio. In other sections results of a deposition of thin films of other high-purity refractory elemental metals like Mo, W, Ti, and Co will be presented as well as the preparation of cast disilicide targets and deposition of thin films of disilicides for diffusion barriers of a high physical quality.

http://dx.doi.org/10.5772/intechopen.69126

2. Production of highly pure refractory targets and thin films

2.1. Purification of refractory metals

Naturally, in order to prepare highly pure sputter targets, the purest initial refractory metals are selected. For example, to produce the purest Ti sponge, the last versions of the Kroll process are included. Then a Ti sponge is elaborated with effective vacuum metallurgy techniques resulting highly pure metal Ti. In general, the basic way of a controlled purification of refractory metals from gas-forming interstitials and metal impurities is high-temperature vacuum melting or annealing. During this process the dissolved atoms of these impurities diffuse to the metal surface and desorb from it. According to modern representations [9, 10], the process of evolution of dissolved gas-forming atoms from the bulk of liquid metals to a vacuum consists of three successive stages: (*a*) a diffusion to the surface of the melt, (*b*) a transition through the interface to an adsorbed state, and (*c*) a surface recombination of adsorbed atoms with a formation and further desorption of diatomic molecules. Stage (*c*) in this chain is essentially nonlinear. Mechanisms of the O evolution depend on the metal nature and consist of the direct desorption of adsorbed O atoms in the atomic state and of the formation and further desorption of metal oxides of different stoichiometries. On the basis of principles of an evaporation deoxidation, the high-temperature behavior of O in refractory metals in vacuum can be roughly divided into three types. Ti, V, and Cr are characterized by a practically complete absence of the evaporation of O and metal oxides, because O in these metals is strongly bound with a metal matrix and has a very high thermal stability. The vapor pressure of the matrix metal (Ti, V, Cr) is much higher than other oxides of these metals. Mo and W are characterized by a high excess of the metal oxide vapor pressure above the metal vapor pressure; therefore dissolved and then chemosorbed O desorbs either independently or as metal oxides. Nb, Ta, Zr, and Hf behave intermediately. It seems that one has an opportunity for a preliminary estimate of the O behavior at the high-temperature vacuum treatment of both solid and liquid metals. A very different situation exists for C atoms which are strongly bound with metal and do not desorb independently. The basis of C evolution to vacuum is an interaction of O and C on the metal surface and further desorption in the form of gaseous CO. Naturally, it is very important to collect theoretical and experimental data on the behavior of such nonmetal impurities as O and C, because this information is a basis for the development of the commercial metallurgical technologies of the production of the high-purity refractory metals for microelectronics. A complex study has been fulfilled which allows one to establish the kinetic connection between the behavior of the mean concentrations of O and C. The dependence of this connection was studied on initial contents of gas-forming interstitials in liquid metals, as well as on the temperature, gaseous phase, and "diffusional transparency" of liquid metals. It has been shown that critical concentrations exist which are typical of each refractory metal. Below this critical concentration, O does not react with C, and the concentration of the latter during vacuum treatment remains constant. Experiments are also made, which demonstrated a high possibility of producing high-purity refractory metals. It has been shown that the "diffusional transparency," when diffusion does not influence chemical processes in liquid metals, can be realized in both vacuum levitation crucible-less melting and electron beam floating zone melting because transport limitations can be removed by the constant renewal of the reaction surface. As for purifying refractory metals from metal impurities, it is important to emphasize that melting temperatures of refractory metals are high enough comparing with ones of many dissolved metal impurities in refractory matrixes. During high-temperature vacuum melting or annealing of refractory metals, dissolved atoms of gas-forming and metallic impurities diffuse to the metal surface and desorb from it. It means that there are optimal conditions for vacuum evaporation of metal impurities because their vapor pressure becomes very high at T_m of refractory metals. Optimal conditions of a vacuum refining have been identified and demonstrated a possibility of the production of Mo, W, Ti, Co, and other refractory metals having low contents of O and C (about 10^{-6} at.%) [11–16]. The opportunity has been demonstrated of the preparation of high-purity refractory metals, when contents of both C and O are at the determination level of the analytical techniques, such as deuteron activation, fast neutron activation, and mass spectrometry with inductively coupled plasma. The relationship between the purity of refractory metals and physical quality of deposited thin films was studied depending on the magnetron and laser sputtering conditions. This is not surprising since only gradually is it realized by scientific community dealing with these materials that the preparation of ultrapure refractory metals might be less difficult than a suitable and thorough trace chemical characterization. It is obvious that the analytic data at the ppb level are difficult to obtain and that interlaboratory comparisons exhibit a scatter of values, which is inversely proportional to the concentration level. This is a natural phenomenon symbolizing the state of the art of the analytical characterization of materials and should not be interpreted as an incompetency of analytical laboratories involved. **Figure 1** gives a rough estimate of the situation experienced by participants in many round robins analyses. However, a scatter of analytical round robin results, especially, on the ppb level of analysis is very hard—discrepancies between results of same samples but at different labs could achieve 100% and more. This is not too surprising that a scientific community has to understand gradually and to

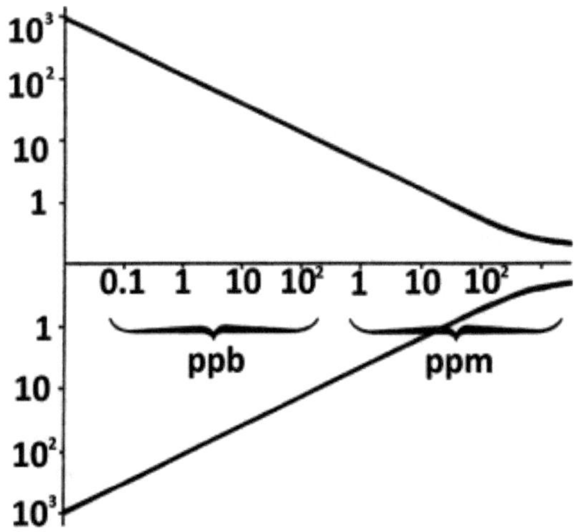

Figure 1. Scatter of round robin results at the trace and ultra-trace levels as a function of concentration levels of trace components.

realize when dealing with these metals that the preparation of ultrapure refractory metals might be less problematic than its elemental characterization [2, 3]. Analytic measurements at the ppm and ppb levels are difficult to obtain, and that interlaboratory comparisons exhibit a scatter of values, which is inversely proportional to a concentration level. This is a natural phenomenon symbolizing the state of the art of the analytical characterization of materials and should not be interpreted as an incompetency of analytical laboratories involved.

2.2. Technologies for production of refractory metals

Three main metallurgical procedures are developed to produce cast refractory metals of high purity for magnetron targets. Procedure #1 is the complex technology for producing elemental metal ingots/targets and bimetallic targets (e.g., Mo/Cu). It consisted of multiple electron beam (EB) vacuum melting of PM compacted blanks of refractory metals, hot pressing, hot rolling, electrochemical etching, electrochemical plating, high-pressure vacuum diffusion welding at high temperatures, machining, etc. Then the bimetallic Mo/Cu targets can be made of the Mo sheet and copper base joined together by vacuum diffusion welding. The upper high-purity Mo sheet of bimetallic Mo/Cu targets serves as a sputter material, whereas the Cu base serves as a heat-conducting material. Similar experiments are successfully done on other metal pairs such as Mo/Cu, W/Cu, Nb/Cu, NiV/Cu, etc. However, the most reliable results are obtained at the production of Mo/Cu. In spite of the lack of a reliable experience of using such targets in commercial sputter setups before, with this technology, hundreds of bimetallic targets are produced for several commercial microelectronics companies in Russia. Procedure #2 consists of multiple EB melting of PM compacted blanks in a vacuum of 1×10^{-6} Torr. Sometimes, the poly- and single-crystalline rods of EB floating zone melting can be used because they are much purer than PM sintered blanks. Two types of water-cooled copper molds, vertical and horizontal, are used for melting and solidification of round ingots of 80–150 mm in diameter and 1.300 mm length and flat ingots of sizes 35×100×1300 mm. Then ingots are carefully deformed and/or machined to get sputter targets. The rectangular and cylindrical molds are used to produce disks or plates of different sizes and diameters (80–210 mm). Procedure #3 is a duplex process, e.g., it consists of multiple EB vacuum melting and/or electric arc vacuum melting [16]. The liquid metal in the mold of the electric arc vacuum setup is intensively stirred by the electromagnetic field of the powerful electromagnetic solenoid. Thus, in this combination, EB vacuum melting is used mainly for a vacuum purification of liquid metals from gas-forming interstitials and metallic impurities, while electric arc vacuum melting is used mainly to produce ingots with a fine-grained macrostructure (this combination of techniques was used mainly for producing W or Mo).

2.3. Sputter of targets and deposition of thin films of refractory metals

Micro-metallurgy aspects imply a study of the magnetron sputtering and deposition of thin films and to study an influence of some parameters of the thin-film metallization on physical properties of deposited metal films. Refractory metal films are deposited on Si(100) wafers of 10–20 Ω cm with/without a thin film (about 0.3 μm) of SiO_2 at room temperature. The Si substrates are cleaned chemically prior to be loaded into the magnetron sputter apparatus. For depositing thin films of metals, magnetron sputter apparatuses of two kinds are used: (*a*) the magnetron sputter system with the planetary arrangement of Si substrates for thin film depositing and (*b*) the conveyer system when Si substrates are deposited during traveling through different chambers of the apparatus. A special care is taken to exclude such contaminants as O, C, and alkaline metals from the apparatus environment. A vacuum at the sputter vessel is about 10^{-6} Torr prior to sputtering. Magnetron targets are cleaned preliminary for 40 min in vacuum. During sputtering, the vessel is filled with Ar of high purity to a pressure of about 10^{-3} Pa. The relative atomic impurity concentration of Ar is less than 3×10^{-6} %. Before sputtering, a heating of Si wafers to 250–300°C is carried out. It has been found that such a procedure is sufficient to produce a clean surface. The deposition rate is practically proportional to the sputtering power for high-purity refractory metals at a constant pressure in a sputtering vessel (**Figure 2**).

As-deposited films are annealed in vacuum. The sputter rate and layer thickness are controlled with the microprocessor and profilometer, respectively. The electrical resistivity of films is measured using a standard four-point probe. Generally speaking, film properties are affected not only by a deposition process as a whole but also by an initial quality of sputter targets. The specific resistivity of cast metal targets and thin films of highly pure refractory metals are shown in **Table 2**. It is well known that the specific resistivity of refractory metals is an integral characteristic of their purity. To study the influence of macro-metallurgical procedures,

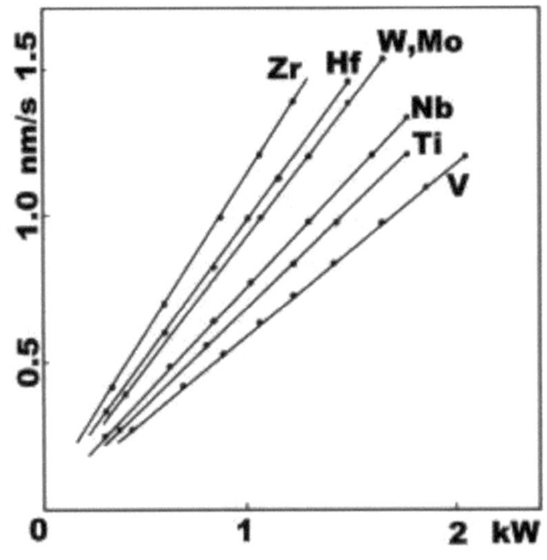

Figure 2. Dependence of a sputter rate for refractory metals on a sputter power at P_{Ar}=1 Pa.

Condition	Metal							
	Mo	W	V	Nb	Ta	Ti	Zr	Hf
Film	5.17	15.0	27.1	17.4	16.8	49.2	41.8	48.3
Target	5.2	5.5	25.0	–	12.7	41.7	–	–

Table 2. Specific resistivity of films and cast metal targets produced of highly pure refractory metals.

used for a production of targets, on the specific resistivity of thin refractory metal films, there are sputtered targets produced both by the standard PM procedure and EB vacuum melting. The content of gas-forming elements (C, O, N) in PM Mo targets is at least 100 times higher than in EBM Mo targets. It should be also mentioned that after prolong vacuum annealing of both targets and thin films, the content of gas-forming interstitials is still very high. The specific resistivity of Mo films of 0.15–1.0 μm thick deposited by sputtering the PM and EBM targets under the same sputtering conditions are 20–35 and 5.17 μΩ cm, respectively.

These results show that an initial purity of the target metal has a very strong influence on the specific resistivity of thin Mo films. After long sputtering, magnetron targets have an erosion path of about 10 mm depth and 20 mm width, and the further sputtering process is characterized by a little bit higher instability. The quantity of sputtered material is about 15–20% of the mass of the target depending on the target design and intensity of a sputtering process. This also shows that the design of targets and magnetrons should be optimal. An effect of a substrate heating on the specific resistivity is studied. When films are deposited on unheated substrates, their electric resistivity is increased by 50–150%, and the scatter of the specific resistivity is greater by 30–40%. This can be accounted for an influence of gas-forming impurities adsorbed on the surface of substrates. A confirmation of this fact is found in the lowering of the specific resistivity of Mo films deposited on unheated substrates with sublayers of Ti or V. These metals are sputtered from targets in the same sputtering vessel and probably react with gases adsorbed on a substrate surface. It is supposed that additional thin layers should not have a strong effect on resistivity measurements. To elucidate the influence of gas-forming impurities on the specific resistivity of refractory metal films, the sputtering power (or deposition rate) is varied and an air is introduced (1×10^{-5}, 4×10^{-3}, 1.3×10^{-4}, and 4×10^{-4} Pa m^3 s^{-1}) during sputtering. Experiments have confirmed that the specific resistivity of high-purity films of Mo, Ti, and Zr depends strongly on the deposition rate and on the presence of reactive gases in the deposition area (**Figure 3**). The ratio of the film resistivity ϱ_v to the target resistivity ϱ_m is a characteristic of the procedure as a whole: the higher the ratio, the less clean is the procedure. In other words, the specific resistivity of the films depends on the ratio of quantities of sputtered (deposited) atoms and interstitials dissolved in the metal films. The dependence of the specific resistivity on interstitials dissolved in the deposited film is studied (**Figure 4**). To analyze this dependence, the atomic concentration C_i of interstitials from reactive gases can be written as $C_i = (1 + \gamma_m S/\gamma_g S)^{-1}$, where γ_m and γ_g are specific rates of the metal deposition and condensation of molecules of n-atomic reactive gases in the refractory metal film, respectively, and S is a deposition area. The rate of dissolution of reactive gasses in the deposited metal film is much higher than the rate of a gas exchange in the deposition area of the magnetron sputtering setup, e.g., for the air leakage Q_i, we had $Q_i = \gamma_g S$. Here, it is assumed that the rate of dissolution of reactive gases in the film is constant. Because the quantity of metal atoms, $\gamma_m S$, which is sputtered in a unit time is nearly proportional to the sputtering power W, we had $\gamma_m S = kW$, where k is the coefficient of proportionality. This coefficient changes slowly with sputter parameters and is determined by dependence of the sputtering coefficient of refractory metals on the energy of sputtering ions and the geometry of the sputtering setup. Considering these equations, we can receive $C_i = (1 + kW/nQ_i)^{-1}$. This equation describes the dependence of interstitial content in the film on the ratio of the sputtering power to the reactive gas flow into the deposition area. The dependence allows us to present curves in **Figure 3** as the dependence of the specific resistivity on the atomic concentration of interstitials in films (**Figure 4**). The air leakage

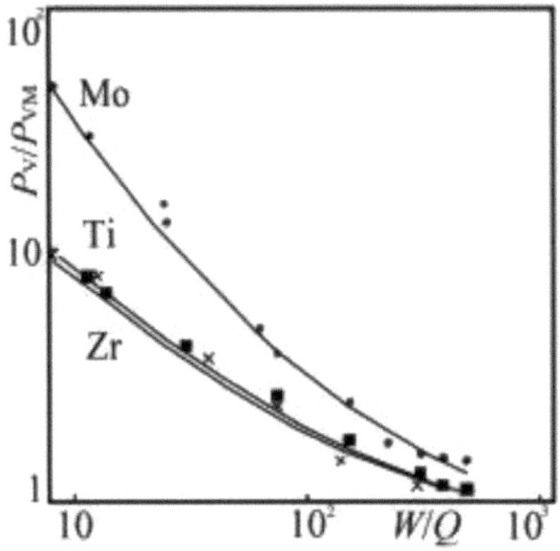

Figure 3. Dependence of film resistivity on sputtering power *W* and current leakage *Q*.

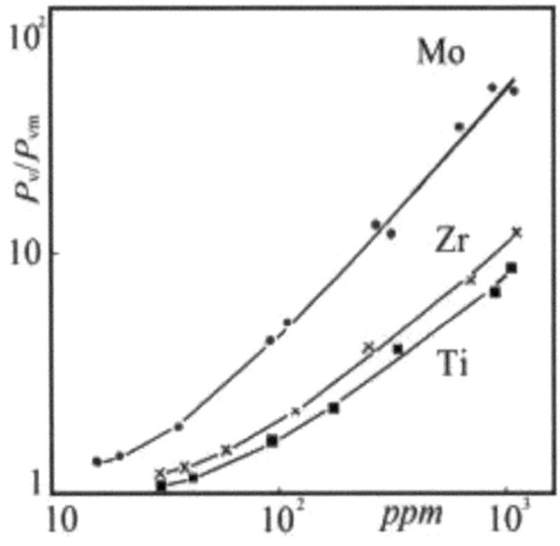

Figure 4. Resistivity ratio as function of interstitial content in the deposited films.

influence becomes very strong as the atomic concentration of interstitials in the film becomes higher than about 10^{-3} at.% (about 10 ppm).

2.4. Depositing molybdenum layers

2.4.1. Short background

Materials and technology of the deposition of thin metal films largely determine a level of performance and reliability of integrated circuits. Increasing the degree of integration and performance of IC necessitates the search for materials and the development of a technology for the deposition of films in addition to or replacing the traditionally used Al (due to the appearance of high-temperature processes) and poly-Si having a high surface resistance. The renewed interest in the use in the IC of refractory metals, e.g., Mo [17, 18], nevertheless, is complicated by difficulties in obtaining films with properties of massive samples. Mo has a rather low specific resistivity and the closest to Si value of the coefficient of thermal expansion. It practically does not interact with SiO_2—the most widely used dielectric in the IC—and has a sufficiently high resistance to a mechanical damage. In addition, Mo forms ohmic contacts with Si with a comparatively low resistance (10^{-5} to 10^{-4} Ω cm^2). In [17], an interesting experiment is done by comparing a high-purity Mo target with a conventional one. It is shown that alkaline metals move easily in gate insulation films and deteriorate properties of MOS interface. The high-purity target which contains no more than 0.01–0.03 ppm Na and K and the conventional one with about 10 ppm of these metals are used in this experiment. The mobile ion quantity in the Mo gate MOS diodes which are produced with either the high-purity target or conventional one are compared before and after the annealing. Results show that no mobile ions are observed in the diode made with the high-purity target. The experiments conducted using W targets show similar results as well. By using several W targets which contained different amounts of Na, it is reported that there is a strong correlation between Na content and the amount of mobile ions in gate electrodes. In [7, 17], it is suggested that α-rays, directly radiated from electrode and interconnecting materials, are the worst factor in the operational reliability of the highly integrated VLSI. The amounts of α-rays radiated from Mo thin films formed by low or high U content targets are <1 and 700 ppb in the high pure targets and conventional targets, respectively; <1 and 280 ppb in the high pure Mo thin films and conventional Mo thin films, respectively; and $<4.2 \times 10^{-5}$ α cm^{-2} h^{-1} and 1.2×10^{-2} α cm^{-2} h^{-1} from the high pure and conventional Mo thin films, respectively. As can be seen from these results, there is a remarkable difference in amounts of α-rays which causes errors. Some decades ago, sputter targets contained several tens ppb and even ppm of U, while now in production scale, targets can often contain less than 1 ppb. EB evaporation makes it possible to obtain films with the specific resistivity ϱ_v = 15 μΩ cm only when the substrates are heated during the deposition to 400°C and higher. Cathode sputtering due to the high reactivity of Mo, low deposition rate (<0.5 nm s^{-1}), and difficulties in providing sufficiently small partial pressures of contaminating gas-forming impurities (O, N, H_2O, etc.) in the area of the discharge led to a relatively high level of the resistivity (higher than 30 μΩ cm). Magnetron sputtering systems that provide deposition rates at the level of 1–2 nm s^{-1} and above make the study of the feasibility of using this method of depositing refractory metal films for the fabrication of IC, in particular with MOS structures. It should be noted that such structures, *ceteris paribus*, are particularly sensitive to impurities such as alkali metals, which diffuse rapidly in the ionized state through dielectric SiO_2 films when voltage fields 10^5 V cm^{-1} are applied. This imposes additional conditions on the parameters of the target material and, of course, on film deposition conditions

2.4.2. *Depositing molybdenum films*

In our study [19] of Mo films, the procedure #1 is mainly used for purification and production of high-purity Mo targets. It is consisted of multiple EB melting of commercial PM bars/

rods at a melting rate of about 0.5 kg min^{-1} in a vacuum of 1×10^{-6} Torr. The power is 250 kW at an accelerating voltage of 25 kV and an emission current of 4 A. Two types of water-cooled copper crystallizers (molds) are used: cylindrical and rectangular. Analysis of the impurity content in Mo targets before sputtering is carried out by means of highly sensitive analytical methods: fast neutron activation, deuteron activation, mass spectrometry with inductively coupled ions, etc. The trace element composition of Mo targets is represented in **Table 1**. Mo films are deposited on Si(100) wafers (10–20 Ω cm) with/without a thin film (about 0.3 μm) of SiO_2 at room temperature. Substrates are cleaned chemically prior to load into a magnetron sputter apparatus. The sputtering chamber pressure is 10^{-6} Torr prior to sputtering; magnetron targets are trained for 40 min in vacuum. During sputtering, the chamber is filled with Ar of high purity to 10^{-3} Torr. Before sputtering, a heating of Si wafers at 250–300°C is carried out. It has been found that such a procedure is sufficient to produce a clean surface. The deposition rate is practically proportional to the sputtering power at a constant pressure in the sputtering chamber. The resistivity of samples of EB-melted Mo is 5.2–5.6 μΩ cm. The specific resistivity of Mo films 0.15–1.0 μm thick deposited from PM targets under same sputtering conditions is 20–35 μΩ cm. These experiments show that an initial purity of target has a very strong influence on the resistivity of thin Mo films. An effect of a substrate heating on the specific resistivity is also studied. An increased scatter of the resistivity is accounted for by the influence of gaseous impurities adsorbed on the substrate surface. To elucidate the influence of gaseous impurities in the specific resistivity of the Mo films, a sputtering power (a deposition rate) is varied, and an air leakage is introduced during sputtering. The experiments have confirmed that the specific resistivity of the films depends strongly on the deposition rate and reactive gases in the deposition area (**Figure 3**). The ratio of the film-specific resistivity (ϱ_v) to the target-specific resistivity (ϱ_{vm}) is a characteristic of the procedure quality: the higher the ratio, the less clean is the procedure. In other words, the specific resistivity of the film depends on the ratio of the quantities of the Mo sputtered (deposited) atoms and interstitials dissolved in the film (**Figure 4**). To obtain ingots free of pores and having a uniform distribution of impurities, a double run is sufficient with a melting rate 0.9–1.0 kg min^{-1}. As a starting material, the rods of commercial purity are used. The samples for the elemental analyses and metallographic studies are cut from the Mo ingot. A composition of trace impurities in the cast Mo under study can be seen in **Table 1**. The macrostructure of cast Mo ingots consists of grains with a length of 40–60 mm and an average diameter of 0.2–3 mm. As substrates for deposition, Si wafers of 76–100 mm in diameter are used, covered with a thermally grown SiO_2 of 0.05–0.3 μm thick. Mo films are deposited in a planar-parallel setup with the magnetron sputtering system. Relative to the target in a plane parallel to the target and situated at a fixed distance from it, substrates are linearly moved on which a metal film is deposited. The setup is continuously moving and equipped with gateways for loading and unloading substrates, providing the setup working without breaking a vacuum. In the setup used, the principle of a vacuum lock of the discharge area is used. As the working gas, a purified Ar is used, which is fed through a leak valve into a discharge area connected to a suction volume of diffusion slots for the passage of conveyor substrates. The pressure in the volume to feed airless Ar is less than 5×10^{-5} Pa and, in the process, less than 2.5×10^{-2} Pa; a discharge current is up to 15 A at a discharge voltage up to 600 V. The unit has an ability to preheat substrates by infrared lamps. Sputtering target composed of four elements is set out in a cooled holder. On each element of the target, O-shaped sputter area (erosion) forms, whose length exceeds the width of the deposition area, thus ensuring the reproducibility of the thickness of the deposited layer on substrates with an accuracy of 2% (**Figure 5**). Ar flow during the deposition is 1×10^{-3} to 2×10^{-2} Pa m^2 s^{-1}. According to approximate estimates, a leakage of gas from the surrounding atmosphere in the discharge area is less than 1×10^{-5} Pa m^3 s^{-1}, which corresponds to the concentration of impurities introduced in the film less than 10^{-4} m.%. When depositing the films, it is revealed that the rate of Mo deposition, depending on the process conditions (Ar pressure, voltage, and discharge current) with accuracy of 10%, is a subject obtained by an analysis of the process. The length of the cathode dark space is determined experimentally by measuring a probe potential distribution in the discharge. In the field of real mode dispersion, it is not more than 0.1 cm, which corresponds to estimates by the formula of Child-Langmuir. Furthermore, the parameter $t = L/l \leq 0.2$ is defined, which greatly facilitates the integration of the equation. Dependence F for Mo is determined for voltages in the range of 300–500 V. The calculated dependence of the deposition rate in a normalized form concerning the conditions, under which there is no scattering of the sputtered Mo atoms on the Ar atoms, is shown in **Figure 6**. Experimental data are obtained by measuring a thickness of the layers deposited

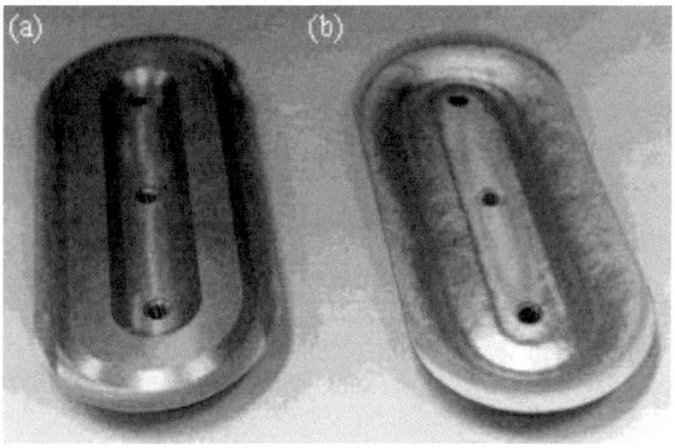

Figure 5. Two magnetron targets: new (a) and (b) after 200 cycles of sputtering.

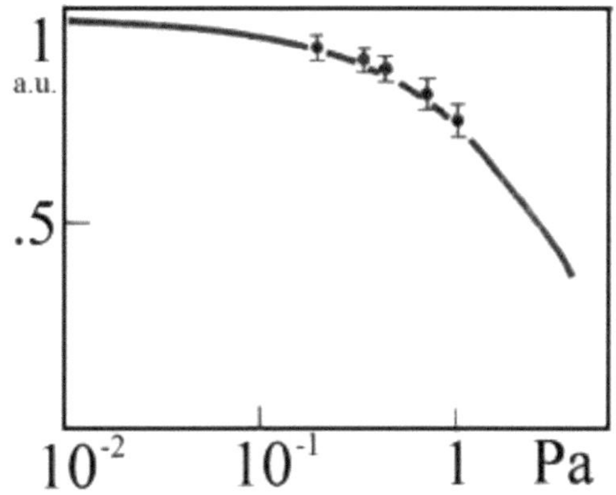

Figure 6. Dependence of deposition rate on Ar pressure: points, experimental data, and curve, calculated data.

under different conditions for specified periods of time. The thickness is measured with a precision micro-interferometer with accuracy of 300 Å. To correctly determine the deposition rate, the layers of 1 mm thickness are used.

To assess the quality of the deposited Mo films, the dependence of the resistivity is determined based on the Mo films on both the deposition conditions and the current-voltage characteristics of the MOS structure. The resistivity is calculated from the results of surface resistance measurements by four-point probe method and a Mo layer thickness with pre-coating a thin layer of Al (0.01 μm) to eliminate a phase distortion. Furthermore, after obtaining the deposition rate depending on the processing mode (the discharge voltage, current, and Ar pressure), the resistivity is determined from the measured surface resistance and the layer thickness calculated from the processing mode. It is found that the resistivity of Mo films in the thickness range of 0.15–1.0 μm with the deposition rate of higher than 1 nm s^{-1} changed a little, and its average value is 10 μΩ cm. Preheating the substrate reduces the resistivity to 8.5 μΩ cm, and it begins to depend on a layer thickness. Apparently, this is due to the influence of residual gases adsorbed on a wafer surface. It is interesting to note that when using PM Mo of a commercial purity (grade M−1) as an initial material for the target, a resistivity of the Mo films with a thickness of 0.5 μm cannot drop below 15–20 μΩ cm. To assess the specific applicability of the Mo films, test structures are fabricated on Si wafers with a resistivity of 7.5 Ω cm with orientation. The cross sections of a MOS test structure including the capacitor, element for determining the contact resistance of Mo-Si, and transistor are prepared. One of them (transistor) is shown in **Figure 7**. The fixed charge density in the oxide and fast surface states near Fermi level are determined on the high-frequency and quasi-static capacitance-voltage characteristics of the capacitor, which are, respectively, 5×10^{10} and 9×10^{10} cm^{-2}. Shifting the flat-band potential after a thermal treatment at 200°C and the field voltage $\pm2\times10^{6}$ V cm^{-1} does not exceed 50 mV for 5 min, which indicates the high stability of the obtained structures.

Measurements of the flat-band voltage on structures are made with a change in the dielectric thickness, depending on the allowed, to determine the difference between the work function of the metal semiconductor, $\Delta\phi_{ms}$, which prove to equal to −0.285±0.015 V. In the case of poly-Si gate, the work function difference is −0.9 V. It is known that a more positive value should significantly improve the current-voltage characteristics of the transistor with a short channel. In order to confirm the feasibility of this assumption, the transistors are

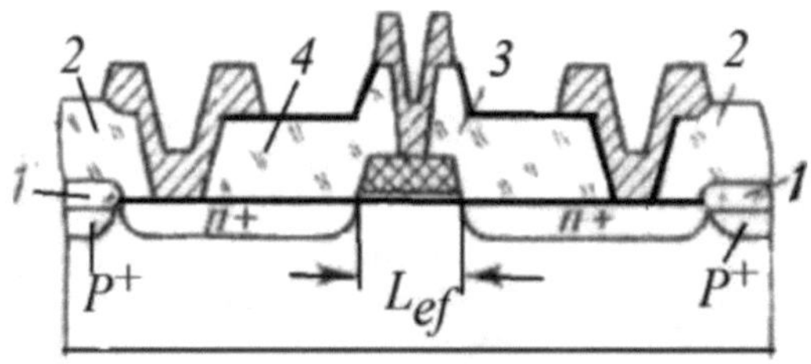

Figure 7. Test MOS structure (transistor); 1, SiO_2; 2, PS glass; 3, Mo; and 4, Mo+Al.

made with an effective channel of a length L_{ef} = 1.2 μm at the channel length L = 2.0 μm on the gate. To eliminate the effects of shortening the channel, ion implantation of the channel by B is made. Measurements show that the magnitudes of the threshold voltage U_t = 1.0 V at the voltage of "source-drain" U_{sd} = −2 V and the "source-substrate" U_{ss} = −2 V are reached on the transistors with gates made of poly-Si and Mo, respectively, at doses of 0.12 and 0.06 a.e. Comparing the current-voltage characteristics shown in **Figure 8** for the transistor of the same geometry, it is revealed that the design efficiency of a poly-Si gate is lower than the structure having a gate made of Mo. This is due to the fact that with increasing doping with B, there is a more severe degradation of mobility in the short channel as well as reducing the slope. Finally, along with the achievement of the resistivity of ~10 μΩ cm of Mo film, which is a very significant for use in IC as an element of their design, the ability has appeared to implement ohmic contacts with Si having a low resistance level, especially in the case of shallow (less than 0.5 μm) transitions. To clarify this possibility and evaluate the stability of the contact, the dependence of the contact resistance on heat treatment conditions is necessary for rapid annealing of surface states. The results (**Figure 9**) show that it remains at $2{\times}10^{-6}$ Ω.cm. It is found that the specific resistivity of the contacts depends on the temperature of the heat treatment as well as the leakage current of n^+-p-transition on the level of ~10^{-8} A cm^{-2} is constant at various temperatures and duration of the heat treatment. Additional studies have confirmed that this is due to the formation of $MoSi_2$ at the boundary of Mo-Si. It should be noted that due to the fine-grained structure of Mo film (grain size of about 700 Å), a photoengraving provided receiving patterns of paths and the size of gaps between paths of about 3 μm and besides the restriction is limiting the possibility of obtaining the necessary quality patterns on a photoresist. The uniformity of the structure and the reproducibility of Mo film properties provided highly reproducible results in a large number of plates are confirmed as well. Thus, the studies show real possibilities and advantages of using highly pure Mo for obtaining VLSI.

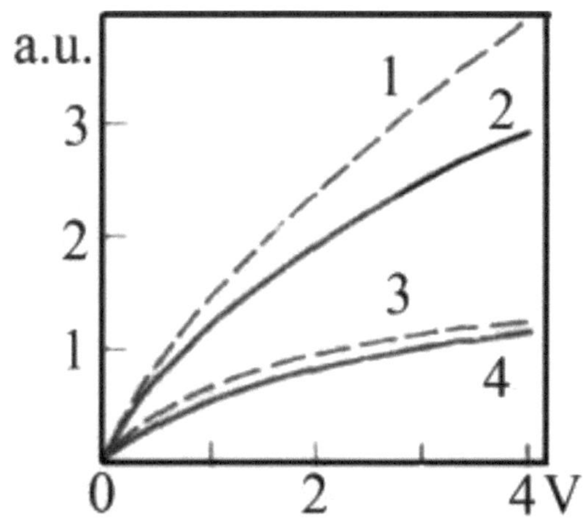

Figure 8. Current-voltage characteristics of transistors with the same geometry of gates produced of Mo (1, 3) and of poly-Si (2, 4); U_g: (1, 2) 4 V, (3, 4) 2 V; Δ_{SiO2}=400 Å; L_{ef}=1.2 μm.

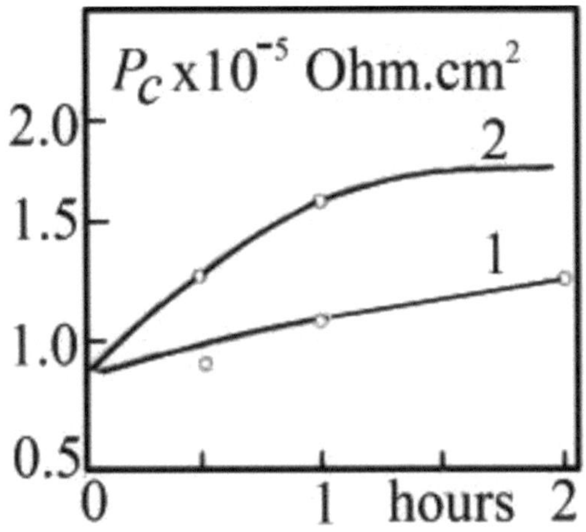

Figure 9. Dependence of contact resistivity on the temperature and duration of heat treatment which were necessary for rapid annealing of the surface states; temperature, (1) 450°C, (2) 475°C.

2.5. Depositing titanium-tungsten layers

2.5.1. Short background

Multilayer structures have become widely used in silicon VLSI technology to produce thin-film current-conducting systems. Normally, they contain contact layers, barrier layers, and base current-conducting layers. In some cases, barrier layers are simultaneously contact-to-Si layers. The reproducibility of the current-conducting system of silicon VLSI as a whole depends to the high extent on physical and chemical properties of diffusion barrier layers between the Si and Al. Naturally, properties of barrier layers connect with the purity of materials, the solubility of other chemical elements in the multilayer structure during thermal processing, interfaces between layers, and manufacturing techniques used. Chemical impurities dissolved in films should not cause any strong radiation damage during the process of film deposition. Current-conducting systems with Ti barrier layers are characterized by the formation of a $TiAl_3$ phase at annealing temperatures above 485°C, which leads to consumption of a barrier material. The solubility of Si in TiAl is several times higher than in Al. Therefore, if barrier layers are not sufficiently thick or contain defects, a degradation rate of the Si/Ti/Al or Si/PtSi/Ti/Al contacts may turn out to be unacceptably high. Films of TiW alloy exhibit a better combination of strength and plasticity than Ti, which considerably reduce a probability that vacancies and point puncture-type defects would be formed in films when residual mechanical stresses are high. It is interesting to study, simultaneously with general conditions of a formation of current-conducting systems, a variety of factors affecting properties of the layered current-conducting systems in VLSI structures. It is found that the content of Ti in TiW barrier layers can be varied in the range 12–43 at.% Ti. At the same time, Ti/W targets, sputtered or evaporated for the deposition of TiW barrier layers, contain no less than 30–40 at.% Ti. The correlation between the Ti content in sputter targets and in deposited thin layers is necessary to develop a controllable deposition process. However, such correlation is very difficult to find and achieve because it depends on many different parameters, such as sputtering conditions, a chemical purity of metal components, a design of the sputter setup, and technologies used for the preparation of targets. To the best of our knowledge, no studies have been reported on using of Ti/W composite cast targets for the deposition of TiW barrier layers. In this chapter, scientific and technological results are presented on both the production of high-purity metals by the vacuum-melting technique and deposition of thin TiW layers by sputtering of composite targets [6, 7, 20–22]. When it is necessary to prepare thin films of alloys, targets composed of the cast metal blocks or co-sputtering of several cast metal targets can be used. As a rule all metallic components should be ultrapure, especially with respect to "mobile ions." With increasing miniaturization, the intrinsic radioactivity also exerts harmful effects; therefore, various components should be "free" of U and Th contamination. Gas-forming interstitials (O, N, H, C) are detrimental too. Because of strict requirements to the purity of refractory metals for microelectronics, all undesired impurities should be removed to less than the level of ppm level in order to be able to prepare extremely pure refractory metals or binary alloys based on such pure metals. Hence, considerable effort is necessary in the purification and doping processes. All process steps should be monitored by very sensitive and, generally, very expensive instrumental analytical methods down to a level even of 1 ppb with a special reference to the gas-forming elements which are extremely difficult to determine in metal materials.

2.5.2. Depositing titanium-tungsten layers by co-sputtering titanium and tungsten targets

The targets of 78 mm diameter and 6 mm thick are used for magnetron co-sputtering [20, 23]. They are machined from polycrystalline ingots of high-purity W and Ti (the metallurgical procedure #2). W and Ti disks for co-sputtering have the same sizes. The films are deposited by magnetron co-sputtering high-purity Ti and W disks which are fixed in water-cooled copper holders of the magnetron sputter setup. The Ar pressure is maintained automatically. Substrates of Si (10–20 Ω cm, 100 mm diameter) placed on a planar substrate holder are spun freely on their axis with 15 rotations per minute and moving in the circular zone which passes above the sputtering targets at a frequency of 30 rotations per minute. The distance between targets and substrates is 60 mm. A shield between them enables the preliminary training of targets and realization of necessary sputtering conditions. The substrates are cleaned chemically prior to load into the magnetron sputtering apparatus. Prior to the deposition, the apparatus is evacuated to 10^{-6} Torr; a leakage of gasses into the deposition area is no higher than 10^{-2} Pa s^{-1}. Before the deposition, the substrates are in situ heated to 250–300°C. A discharge power on each target is up to 2 kW; Ar pressure during a co-sputtering procedure is 10^{-3} Torr. Based on an assumption that the deposited layer density does not differ from that of a bulk metal, in order to obtain a necessary Ti/W ratio in the film, deposition rates of both metals are determined as a function of the discharge power at Ar pressure of 1 Pa. For that purpose metal films are deposited on thermally oxidized Si substrates at various discharge powers, the energy consumption during these processes being 400–600 kJ. This ensures the layer thickness of 0.5 μm and the possibility of their measurement with sufficient accuracy. Experiments show that the deposition rate of each metal depends linearly on the discharge power. **Figure 10** depicts dependences of mean and instant rates with respect to the thickness (γ_1 and γ_2, nm s^{-1}) and dependences of mean and instant deposition rates (γ_1 and γ_2, atoms cm^2) on the discharge power. Six groups of film samples are prepared which differ from each other by quantities of monoatomic layers of Ti and W and by Ti/W ratios. Quantities of monoatomic layers of each metal in these six film specimens are shown in **Table 3**. Bearing in mind that the 1-mm-thick film area of 1 cm^2 contains 5.67×10^{15} Ti atoms and using experimental data of **Figure 11**, the sputtering discharge power for each target can be determined which is necessary to prepare homogeneous TiW films with definite but different Ti/W ratios. Because Ti layers alternated with W layers in the film structure, some lamination or layering of the film is possible. To escape this, vacuum annealing is done to make the structure of as-deposited films more homogeneous, which is confirmed by X-ray diffraction. The mean grain size of films with different chemical compositions measured by transmission electron microscopy is in the range 10–20 nm. The structure of samples 2 to 4 revealed by X-ray diffraction (**Table 3** and **Figure 11**) is found to be a solid solution of Ti in *bcc* α-W with a lattice parameter from 0.318 to 0.323 nm (by 0.5–2% larger than that for pure W films—*bcc* β-W has a lattice parameter $a = 0.317$ nm). The structure of pure Ti films is *hcp* α-Ti with $a = 0.295$ nm and $c = 0.447$ nm. Such a surprising result is supposed to be a consequence of intermixing of Ti and W atoms with high energies (1–3 eV), the mutual interdiffusion and an overlapping deposition zones during magnetron co-sputtering although the thickness of deposited layers is about 4 nm.

The specific resistivity of TiW thin films depending on the Ti/W ratio is studied using values of the surface resistivity and the film thickness (**Figure 11**). The resistivity of thin films of

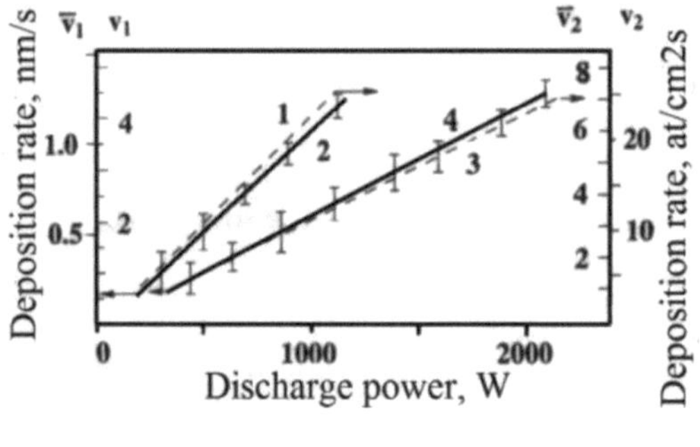

Figure 10. Mean and instant deposition rates as function of the film thickness.

Sample	Monoatomic layers	
	Ti	W
1	1.4	13.8
2	5.0	13.8
3	12.0	14.0
4	11.0	5.5
5	11.0	3.0
6	14.4	1.0

Table 3. Quantities of monoatomic layers of Ti and W in six film samples.

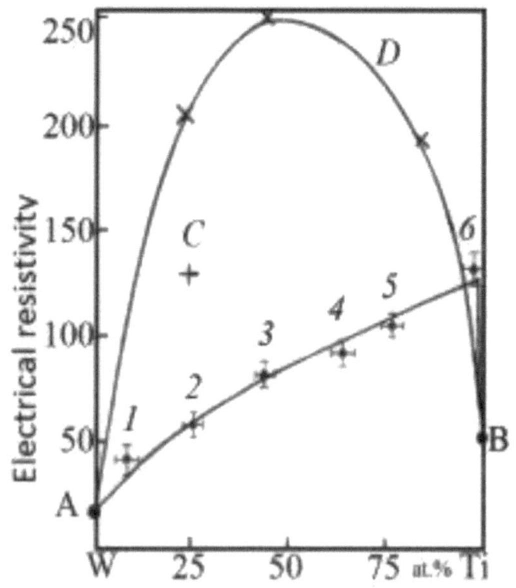

Figure 11. Dependence of resistivity of Ti/W films on Ti/W ratio. Points 1–6 [20], points of curve *D* [22]; point *C* for film deposited from PM target.

pure W and Ti is very near to the tabulated ones for bulk metals. They are shown in **Figure 11** as points *A* for W and *B* for Ti. The coincidence of the resistivity for films and bulk samples can be a characteristic of the quality of the co-sputtering procedure. The resistivity of films is increased gradually as the Ti/W ratio is changed from those for pure W to pure Ti. However, it is worth to mention that the resistivities of quasi-alloy films are always higher than those of pure metals. Such behavior of the resistivity of both pure metals and alloys is well known in physical metallurgy and can be explained by using physical models for a conduction of electrons in solids. From the physical point of view, the Ti/W quasi-alloy consists of two very different metals which have very different physical properties (densities, melting points, etc.) and phase structures (*hcp* and *bcc*). Thus, it is very difficult to make a reliable prediction of the resistivity curve behavior of these films depending on their chemical compositions. Experimental studies of physical characteristics of these quasi-alloys are of a great interest because of the necessity of getting the physical description of the composition, structure, etc. Another experimental fact which is also very surprising is the resistivity of TiW films with a small addition of W, as a dopant (1–4 at.%): the resistivity of a doped Ti became higher than the specific resistivity of pure Ti films. Contrary to this fact, the difference between the resistivities of both pure W films and TiW films with a small addition of Ti as a doping element increases gradually. At the moment there is no a reliable explanation of this interesting fact. Comparing our data with those of Babcock and Tu [22], values of the specific resistivity of thin films deposited by co-sputtering of high-purity cast metal targets are much lower than those of [22]. The discrepancies between the results of these two studies are evidently the consequence of using cast metal targets of higher purity in our study [20] as well as a more advanced deposition procedure. Another result, which is also of some interest, is the difference between the resistivity of films deposited from two kinds of sputtering targets: cast metal targets and PM Ti/W powder targets. By sputtering of targets compacted from high-purity Ti and W powders with W content of 25 at.% by PM techniques, vacuum-annealed films have the specific resistivity of 130 μΩ cm (Point *C*, **Figure 11**). Auger spectrometry of films which are obtained by magnetron sputtering of PM targets shows the presence of high contents of C and O, 5 and 4 at.%, respectively. In layers obtained by co-sputtering of cast metal targets, C and O contents are lower than 1 at.%.

2.5.3. Composite titanium-tungsten targets for deposing barrier layers

Metallurgical technologies to produce of Ti/W targets are selected according to procedure #1 [24]. Multiple EB vacuum melting of compacted Ti blanks is used, as well as cold/warm/hot rolling, cutting, drilling, pressing, etching, polishing, etc. By this procedure, cast Ti polycrystalline disks of 190 mm in diameter and 23 mm thick are prepared. For this purpose, Ti rods produced by iodide process are preliminary carefully purified by EB floating zone vacuum melting and then remelted by EB vacuum melting in specially designed water-cooled copper molds. At the same time, highly pure W single-crystalline rods of 11 mm in diameter and 350 mm length are purified by EB floating zone vacuum melting. Composite cast Ti/W targets consisted of Ti disks with W cylindrical attachments, arranged in such way that the ratio between the areas of Ti and W sections on sputtered targets surface corresponds to the necessary Ti/W ratio in deposited films (**Figure 12**). TiW films are deposited by sputtering composite Ti/W, contained 40 at% of Ti, in a sputter deposition system with d.c. power of 2.4–3.2 kW. Leakage currents are measured at a reverse bias voltage of 15 V. The current density during sputtering is about 0.1 A cm^{-2}. The sputtering yields of Ti and W are the same; however, they are chosen to obtain the necessary Ti/W ratio during the whole sputtering procedure. The deposition rate is about 1.8 nm s^{-1}. Films are deposited on *n*-Si(111) substrates (100 mm in diameter) preheated in vacuum by IR lamps to 250°C with accuracy of 5°C. Test Schottky diodes with both *n*-Si/PtSi/TiW/Al and *n*-Si/PtSi/Mo/Al structures are formed by standard techniques. These techniques consist of (*a*) deposition of thin Pt layers by cathode sputtering on Si substrates through contact windows of 10 μm × 15 μm which are formed in a surface thermal SiO_2 layer, (*b*) annealing of the PtSi structure and removal of the unreacted Pt, (*c*) magnetron sputtering of TiW or Mo, and (*d*) deposition of the upper Al layer by d.c. magnetron sputtering of A1-1 at.% Si target. Photolithography and etching are used to obtain contact windows. The structures are annealed in a nitrogen atmosphere at 450°C for different times, but not more than 3 h. The temperature of Si substrates is measured indirectly by a calibration

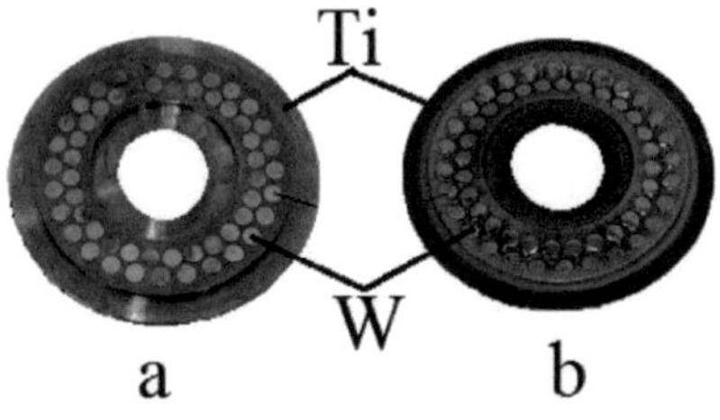

Figure 12. Composite sputter Ti/W targets (left, new; right, after 150 cycles).

or thermal painting. During measurements of mechanical stresses, the heating of all samples is carried out in the same way. The thickness of TiW films is 0.18 μm with accuracy of 0.02 μm. The sheet resistance is 3.5–4.5 Ω/□; the specific resistivity of TiW films is 65–72 μΩ cm. The samples are analyzed by four-point probe sheet resistance measurements, scanning electron microscopy, Auger electron spectroscopy, X-ray diffraction, X-ray electron probe spectroscopy, and laser ellipsometry, enabling an evaluation of residual mechanical stresses in films to be made with error of 2%. The leakage currents and Schottky barrier heights are registered with accuracy of 0.5 and 0.1% over the measurement range, respectively. XRD data show that TiW films are solid solutions of Ti in a W matrix with increased W lattice parameters. A relative increase in the W lattice parameter depends on a quality of surface conditions (polished or unpolished) of Si substrates on which films are deposited (**Figure 13**). The relative increase in the W lattice parameter for a polished surface was $\delta d/d0 = 7\times10^{-3}$, while for an unpolished surface, it is $\delta d/d0 = 2\times10^{-3}$. Here $d0$ is the lattice parameter of W (from ASTM tables). This difference is mainly due to the action of residual mechanical stresses in films. Measurements of a curvature of Si substrates with WTi films show that bending radii were $R_{1b} = 47.0$ m and $R_{2b} = 64.5$ m for polished Si substrates with/without a thermal SiO_2 sublayer on the Si surface. Bending radii are $R_{1b}' = 69.0$ m and $R_{2b}' = 103.0$ m for unpolished Si substrates with/without thermal SiO_2 sublayer. Mechanical stresses of specimens with/without SiO_2 sublayer correlate with a microstructure of TiW films. Mean grain sizes of TiW films are 35 and 45 nm for samples with/without SiO_2 sublayer, respectively. **Figure 13** shows microstructures of surfaces of TiW films on the polished Si substrate with (*b*) and without (*a*) the thermal SiO_2 sublayer.

The bending displacement is reciprocally related to the bending radius of the substrates, and so it can be supposed that mechanical stresses in films deposited on unpolished substrates are lower than those in films deposited on polished substrates. This correlates with values of relative deviations of the lattice parameter for unpolished and polished substrates ($\delta d/d0 = 2\times10^{-3}$ and 7×10^{-3}, respectively). The absolute value of the bending displacement can be used to estimate resulting mechanical stresses in the sample. Values of bending radii for samples without SiO_2 sublayers are higher than for samples with SiO_2 sublayers on both unpolished and polished Si substrates. These results confirm the observations of the dependence of mechanical stresses in thin layers on properties of the material and its crystallographic microstructure. It can also be supposed that SiO_2 sublayers change conditions of nucleation and growth of TiW films. As a result, there is a more spontaneous formation of nuclei, faster unity, and higher resulting stresses localized on grain boundaries. The influence of sublayers on the layer

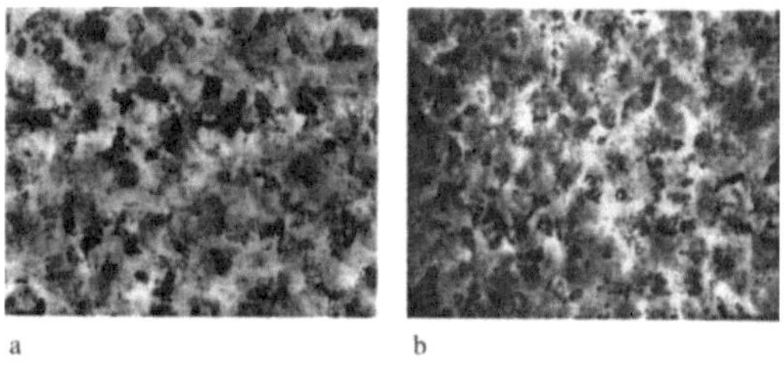

Figure 13. Scanning electron micrographs of the surface morphology of samples with structure Si/TiW (polished Si substrate) (a) and Si/SiO_2/TiW (b) (×100,000).

microstructure is confirmed by experiments in which Al layers are deposited onto both Si substrates and Si substrates with TiW film. The influence of a sublayer material on the film microstructure is studied on Al layers, prepared by magnetron sputtering of Al targets doped with Si (1 at.%), onto Si substrates with/without TiW sublayer. **Figure 14** shows a microrelief of Al films deposited on Si substrates without (*a*) and with (*b*) the TiW sublayer. The microstructure of the Al film in the Si/Al structure is rather smooth, and the Al film consists of coarse grains with a mean grain size of 1.1–1.3 μm. The mean grain size of Al films in the Si/TiW/Al structure is lower (about 0.3 μm) than in the Si/Al structure. TiW films, deposited by magnetron sputtering of composite cast TiW targets, contained approximately 40 at.% Ti, with a remainder W. Because TiW films serve simultaneously as current-conducting and barrier layers, sputtering targets should have a relatively high W content, together with the necessary Ti/W ratio for both optimal barrier properties and lower electric resistivity of TiW films. For effective barrier properties, it is preferable to have the Ti content in a range 30–40 at.% Ti in TiW thin films.

As shown earlier, the specific resistivity of TiW films increases with decreasing the W content to 98 at.%: the specific resistivity for this alloy is approximately 2.5 times higher than that for pure Ti films. The difference in the specific resistivity in the range of 30–40 at.% Ti is not large: 60–70 μΩ cm. To meet the requirements of the composite target design, the Ti content of 40 at.% is chosen. An excellent correlation between Ti/W ratios of targets and of films during the whole lifetime of sputtering targets is found, despite the fact that sizes of the erosion area changed significantly. The contents of the Ti and W components in the films vary slightly after 130–150 sputtering cycles. The relative deviation of the Ti content from the standard one during the target lifetime does not exceed 2 rel.%. Our AES studies of TiW films have revealed low concentrations of interstitials, such as O, C, and P. Auger profiling data of unannealed specimens are shown in **Figure 15(a)**. A typical result for these films is a sharp decrease of O and C contents during the etching process and a deepening of the analyzing zone. The behavior of O and C is very similar, and their curves partially overlap. Mechanical properties and microstructure of TiW layers also depend on the content of gas-forming interstitials, e.g., O, C, and H, even when they are present in very small quantities. It is unclear how strong the influence of a low content of interstitials is on physical properties

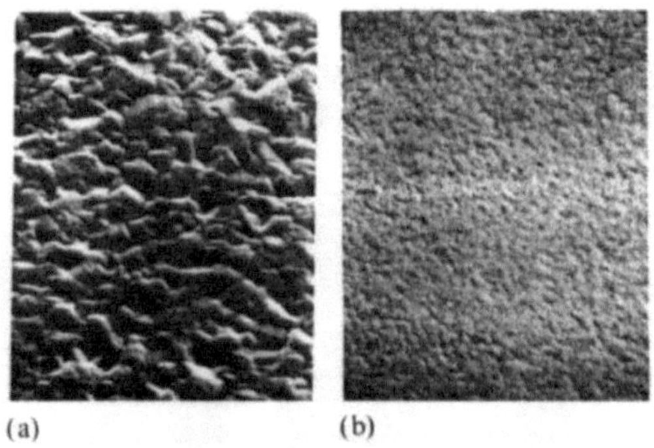

Figure 14. Scanning micrographs of the structure morphology of samples with structures Si/Al (a) and Si/TiW/Al (b) ×100,000.

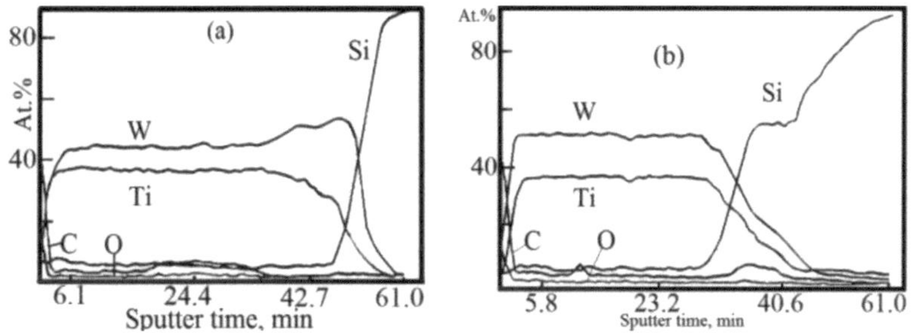

Figure 15. Auger electron spectra of samples with structure Si/TiW: (a) Auger electron spectra of samples with structure Si/TiW before (a) and after (b) isothermal annealing at 510°C for 1 h.

of thin films, although it is well known that interstitials, even at low concentrations, form very stable fine chemical compounds like refractory metal carbides, oxides, nitrides, etc., which precipitate on boundaries in thin metal films and can serve as boundary diffusion stoppers. On the other hand, the lack of discontinuities on the curve for C at an etching time of 50 min, e.g., near the Si/TiW interface, confirms both the effectiveness of the preparation of Si substrate surfaces before deposition and the sufficient cleanliness of sputter and deposition processes, at least at initial stages. A slight monotonic decrease in the Ti content near the Si/TiW interface is probably connected with an increase in the W content near the interface. It may also be a result of the nucleation of TiW films, which is preferable for W atoms because TiW films are based on the W lattice. It may also be associated with the fact that the coefficient of a W self-diffusion is much higher than the coefficient of a Ti self-diffusion (1.88×10^{-8} and 6.4×10^{-8} cm^2 s^{-1}, respectively). This may promote an easier motion of W atoms on a nucleation front during the film growth, as well as more frequent collisions of W atoms with a nuclei, and hence a higher probability of their accumulation on the nuclei. As can be seen from **Figure 15(a)**, W and Ti are homogeneously distributed throughout the film depth, which is probably a consequence of their interdiffusion. Near Si/TiW interface, there is a rather extended silicide layer, as can be seen from a monotonic change in the Si curve. Some localization of C near Si/TiW interface can be related to the differences in a C limited solubility and a C diffusion coefficient between the initial TiW film and the transition silicide layer detected near the SiO_2/TiW interface. The considerable O content in TiW layer is dependent on the fast O diffusion, compared with the self-diffusion of components of TiW alloy, and the high O solubility in W and Ti as well as in their silicides. If no Ti_xW_y is present in a solid solution, it is possible that Ti may display deoxidizing abilities and react with the Si recovered from SiO_2. In this case, a nonstoichiometric Ti silicide phase should be formed, because the formation of the stoichiometric $TiSi_2$ phase is not observed at such low temperatures. This supposition is in agreement with a monotonic change in the Si curve which correlates with a similar O curve. A temperature effect results in the depth redistribution of chemical components and impurities of TiW film. **Figure 15(b)** shows the Auger profiles for the depth distribution of elements in TiW films, isothermally annealed at 510°C in vacuum for 1 h. The W and Ti are homogeneously distributed throughout the film depth which is probably a consequence of the interdiffusion. At the Si/TiW interface, there is a rather extended silicide layer as can be seen from a monotonic change of the curve for Si. An increase in the C content at Si/TiW interface can be related to differences in the C limited solubility and the C diffusion coefficient between initial TiW film and the transition silicide layer.

TiW films with a junction depth of 0.18 μm are tested as barrier layers on diodes with a square shape (10 μm × 15 μm). Schottky barrier heights on *n*-Si are determined from the current *vs.* applied voltage measurements. TiW layers form a relatively low Schottky barrier contacts to *n*-type Si (no higher than 0.5–0.6 eV). Curves 1 and 2 in **Figure 16** show the dependence of leakage currents on the testing time at 450°C in nitrogen. For comparison, barrier layers of Mo are also tested (**Figure 16**, curve 3). The measurements of the direct breakdown potential (U_0) for structures with TiW layers have revealed their high thermal stability after testing at 450°C for 3 h in nitrogen. A similar testing of structures with Mo layers shows that they are of a lower stability than structures with TiW layers (**Figure 16**, curve 3). Our experimental data on the testing of TiW barrier layers are associated with both the microstructure and morphology of Si-Me interface in contact with barrier layers of TiW or Mo. It is worth noting that leakage currents can be lowered by optimizing the technological procedures, for example, by a choice of a sputtering power (**Figure 17**). A higher quality of contacts with TiW barrier

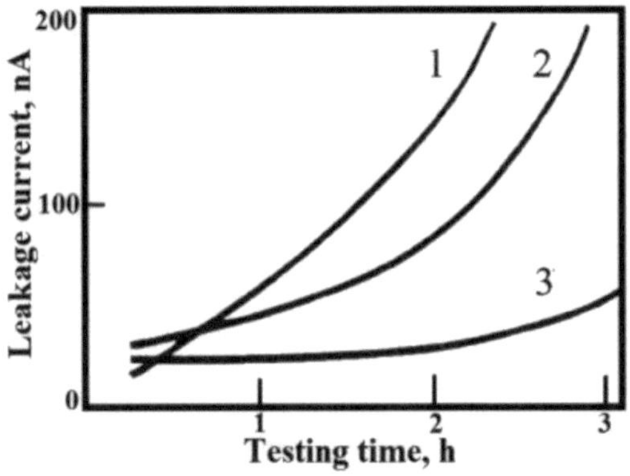

Figure 16. Leakage currents of samples with Schottky barrier diode structures *n*-Si/PtSi/TiW/AI and *n*-Si/PtSi/Mo/Al tested at 450°C in nitrogen: (1, 2) TiW films deposited at sputter deposition powers of 2.4 and 3.2 kW, respectively; (3) Mo film.

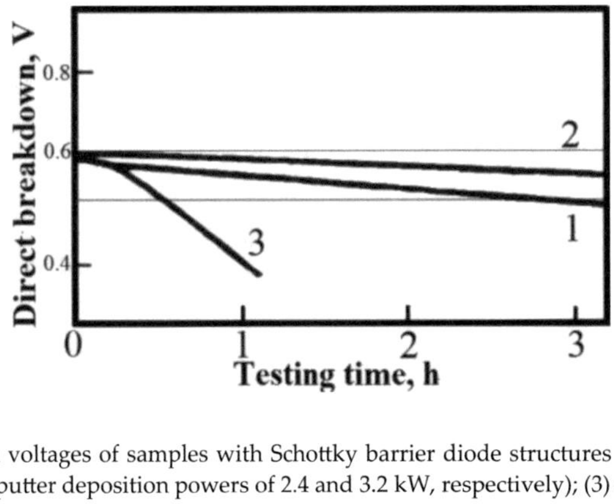

Figure 17. Direct breakdown voltages of samples with Schottky barrier diode structures at a current of 300 pA: (1, 2) structure *n*-Si/PtSi/TiW/Al (sputter deposition powers of 2.4 and 3.2 kW, respectively); (3) structure *n*-Si/PtSi/Mo/Al.

layers, obtained at an increased sputtering power, is partly due to a higher dispersion of the microstructure. Schottky diodes with TiW films exhibit a higher level of leakage currents than diodes with Mo films, although in both cases leakage currents are not an obstacle to the commercial use of these highly pure refractory metals as Schottky diode barrier layers. On the other hand, a great advantage of Schottky diodes with the Si/PtSi/TiW/Al structure is a rather high thermal stability of a direct breakdown potential U_0). A maximum deviation of U_0 from the mean value for structures with TiW layers is no higher than 0.1 V during thermal testing at 450°C for 3 h in nitrogen (**Figure 17**, curves 1 and 2). At the same time, the maximum deviation of U_0 from the mean value for structures with Mo layers is twice as high after a thermal testing at 450°C for 1 h in nitrogen (**Figure 17**, curve 3).

2.6. Depositing films of refractory metal silicides for barrier layers

2.6.1. Depositing films of refractory metal disilicides

Refractory metal silicides have a large potential as materials for the low-resistance contacts, gate electrodes, and interconnections in microelectronic devices [24, 25]. One of the well-known techniques for preparation of silicide layers involves a laser co-evaporation and magnetron co-sputtering of pure metal films onto a surface of a Si substrate with a subsequent high-temperature annealing. Another is magnetron sputtering or laser evaporation of silicide targets produced by PM or vacuum melting/casting. Casting of silicide ingots of commercial sizes is a well-known metallurgical problem because all refractory metal silicides are very brittle and hard in mechanical working. The optimal method is production of composite cast refractory metal silicide targets (cast silicide pieces are attached to copper bases by ultrasonic soldering). Vacuum melting and casting refractory metal silicides for targets solve the two main problems of thin-film deposition, i.e., desired chemical composition and high purity of thin films. The most problematic is a production of cast WSi_2 targets for the laser ablation or magnetron sputter because it is necessary to use two vacuum techniques—vacuum HF levitation and EB float zone melting. Rectangular cast WSi_2 targets used in this study [26] for laser ablation have a volume of 1.5 cm^3 and are chemically homogeneous. For magnetron sputtering composite, cast silicide targets of 152 mm in diameter are produced. Our study has revealed possibilities of depositing WSi_2 films by both the laser evaporation and magnetron sputtering. Other refractory metal silicides are produced in a similar way. Three experimental series are conducted to obtain silicide films: (1) laser evaporation of cast silicide targets, (2) magnetron sputtering cast silicide targets, and (3) magnetron co-sputtering of metal and Si targets. Initial materials for silicide sintering are 5N-purity Si and high-purity refractory metals (**Table 1**). Homogeneous mixtures of high-purity Si and refractory metal are prepared and isostatically pressed. Then mixture samples are sintered in a vacuum at 1100°C for 3 h. PM sintered samples are melted by vacuum HF levitation and cast in copper molds. The levitation melting consists of suspending a solid sample in HF electromagnetic field and melting it by the induced electrical currents. The resulting 25 g liquid drops have a form of the sphere. To solidify a liquid drop, it is enough to switch off an electric power. During laser ablation, the 10×15×5 mm^3 cast targets are evaporated with a solid-state Nd-laser of 1.06 μm wavelength, 0.8 J pulse energy, and 10–20 Hz pulse frequency [26]. The laser beam spot is 2–3 mm in diameter.

The substrates are Si(100), Si(111), MgO, ZrO_2, and Al_2O_3. The substrate temperature is varied within the range of 100–750°C. The deposition time is 5–20 min and a film thickness 15–200 nm. Magnetron sputtering involves the sputter cleaning of cast targets and sputtering films in Ar. Magnetron co-sputtering is as well used to produce many other refractory metal silicides (Ti-Si, Zr-Si, Hf-Si, V-Si, Nb-Si, Ta-Si, Mo-Si, WSi_2, $CoSi_2$, etc.). The sputtering system is first evacuated down to a vacuum of 2×10^{-6} Torr prior to the deposition of thin silicide films. Si wafers are cleaned chemically prior to load into a magnetron sputter apparatus. A special care is taken to exclude such contaminants as O, C, and alkaline metals from an apparatus environment. The sputtering procedure involves sputter cleaning cast silicide targets in Ar for 5–10 min, while the shutter is closed and sputters depositing silicide films in Ar. The relative atomic impurity content in Ar is less than 3×10^{-6}. The resistivity of thin disilicide films as well as cast disilicide targets is shown in **Table 4**.

2.6.2. Depositing films of WSi_2

During magnetron sputtering, the WSi_2 films are deposited onto Si(100) wafers (10–20 Ω cm) with/without a thin film (about 0.3 μm) of SiO_2 at room temperature. The WSi_2 films are deposited by magnetron sputtering cast targets in a sputter deposition system with a constant vacuum of 2×10^{-2} Torr and d.c. power of 0.26 kW [27]. To produce targets for magnetron sputtering, the WSi_2

Disilicide	**Electrical resistivity (μΩ cm)**	
	Target	**Film**
$TiSi_2$	16.9	13–17
$ZrSi_2$	75.8	40–43
$HfSi_2$	62.0	150–260
VSi_2	66.5	67–80
$NbSi_2$	50.4	55–63
$TaSi_2$	46.1	60
$MoSi_2$	21.6	67–80
WSi_2	80.1	50–70
$CoSi_2$	16.0	30.0

Table 4. Specific resistivity of thin disilicide films of refractory metals.

Figure 18. Composite WSi_2 targets for magnetron sputtering: left, new; right, after 150 cycles.

cast blocks of 20×15×5 mm^3 are machined and soldered ultrasonically to copper bases of 152 mm diameter (**Figure 18**). Considering that our purpose is to replace a standard Al metallization with WSi_2 as well as to form the next metallization level, it is not our aim to study a resistivity of contacts of WSi_2 to the *p*-type conductivity layers. Thus, the simplified technological method is used. Following sputtering, the wafers are cleaned in a dilute HF solution for 4 min and rinsed in deionized water before loading into the deposition chamber. The wafers are treated in a buffer etchant for 10 s. A poly-Si layer 0.11 μm thick is deposited onto two wafers. The 0.25 μm thick WSi_2 layer together with a satellite is deposited, and the Al commutation is formed. The WSi_2 layer is reactively ion etched. For removal of the photoresist, a plasma-chemical treatment is employed with additional chemical treatment. Then the wafers are covered with a 0.7 μm borophospho-silicate-glass (BPSG) layer and annealed in N_2 gas. The annealing chamber is purged with N_2 for 10 min before each anneal, and the annealed wafers are allowed to cool in N_2 before removing from the chamber. The BPSG layer is etched off with different chemical procedures.

Various chemical treatments of etching are used on different parts of wafers. Treatment #1 consisted of stripping of the BPSG layer in a buffer etchant for 2 min. In treatment #2, the BPSG layer is stripped off in a similar buffer etchant; however, the layer is etched for 4 min. Treatment #3 involves reactive ion etching of contacts. Treatment #4 includes all three treatments listed above. The input power to sputter composite mosaic targets is as much as 260 W for 30 min for each cycle. Before sputtering, Si wafers are heated to 250–300°C. It has been found that such a procedure is sufficient to produce a clean Si surface. The deposition rate is proportional to the sputtering power at a constant pressure in the sputtering chamber. The sputtering rate and the layer thickness are monitored during deposition with the microprocessor and profilometer, respectively. The deposited silicide films are annealed in vacuum. Some samples are annealed at 950°C in N_2 for 30 min. The film thickness is found to be 200–250 nm. The specific resistivity of cast specimens and of conducting WSi_2 paths as well as contacts to poly-Si layers is measured by the four-point probe sheet resistance method. X-ray phase analysis of cast samples is performed on a Siemens D-500 diffractometer ($\psi = 25°$, Fe-07), and a JXA-5 apparatus is used for local X-ray spectral analysis. Chemical reactions between the film and wafer as well as the distribution of elements are studied by Auger electron spectroscopy together with ion sputtering with a JAMP-10S Auger electron spectrometer. The energy of the primary ion beam is 10 keV, and the current was 5×10^{-6} A. The spot is increased up to 100 μm diameter to prevent a decomposition of the material studied and to lower an electron current density. 1 keV Ar ions are used for sputtering. Instability of an ion gun is no more than 10%. The ion beam diameter is 10 mm. A scanning electron microscope study is performed with a JSM-35S and a Stereoscan 240. Both the elemental composition and elemental depth distribution of 100 nm thick WSi_2 films are analyzed by means of Rutherford backscattering of 1.5 MeV He ions. After annealing of thin refractory metal films deposited in vacuum by magnetron co-sputtering of cast high-purity refractory metals on Si or Si-on-sapphire substrates at 650–700°C, stoichiometric thin disilicide films are produced, which are confirmed by Auger electron spectrometry.

The specific resistivity of cast WSi_2 targets is 50 μΩ cm. It is lower than that of PM targets (70 μΩ cm). Local X-ray spectral analyses of cast targets show that samples are homogeneous. An analysis of the phase composition of laser-deposited films obtained by evaporation (ablation) of cast WSi_2 targets on (100) and (111) Si substrates shows that amorphous WSi_2 forms at substrate temperatures up to 400°C. Above this temperature it crystallizes and exhibits a mixture of the tetragonal phase and a small amount of the semiconducting hexagonal phase. As the substrate temperature is increased, the amount of the hexagonal phase decreases. At 700°C the film is single-phase tetragonal having a metallic type of conduction. At 750°C, a small amount of a second phase (W_5Si_3) reappears. The experimental results show that, as the pulse frequency is lowered, WSi_2 films become multiphase. In this case the main tetragonal phase of WSi_2 is observed together with a small amount of the hexagonal phase and W_5Si_3. The occurrence of a phase deficient in Si with respect to the disilicide composition might be attributed to the evaporation of the latter. The phase compositions of films obtained after annealing and those deposited immediately onto a hot substrate are similar. Interplanar distances for lines on diffraction patterns of sintered and cast WSi_2 targets and of WSi_2 films deposited from cast targets at 700°C are given in **Table 5**. Phase compositions of films are seen to correspond to those of targets. A small deviation of intensities of measured lines from tabulated ones indicates the presence of texture in samples. Rutherford backscattering data indicate that 100 nm WSi_2 films (the Si/W ratio being approximately 2) are obtained on Si substrates (**Figure 19**). An analysis of the elemental depth distribution shows that this ratio remains constant throughout the film depth. No transitional layer between the film and the substrate (Si or neutral) is observed.

In every case, the tetragonal WSi_2, which coexisted with a small amount of W_5Si_3, is the dominant phase. The microstructural studies of the films deposited on Si(111) at 700°C in a vacuum show that they are homogeneous without ruptures. Being rather thin (30 nm), the films probably copy a substrate surface relief. The SEM studies of the cross sections of the films show

hkl	*d*				
	File data	**PM powder target**	**Cast target**	**Film (laser)**	**Film (magnetron)**
002	3.908	3.911	3.907	3.928	3.927
101	2.970	2.970	2.970	2.962	2.970
110	2.270	2.270	2.268	2.263	2.269
103	2.020	2.025	2.026	2.025	2.025
004	1.961	1.963	1.964	1.965	–
112	1.961	1.963	1.964	1.965	1.964
200	1.603	1.606	1.606	1.601	1.608
114	1.485	1.485	1.484	1.483	1.484
105	1.412	1.412	1.413	–	1.410
211	1.407	1/407	1.407	1.409	–
006	1.304	1.305	1.304	1.308	–
213	1.258	1.258	1.258	1.255	1.257
204	1.241	1.241	1.240	–	–

Table 5. Interplanar distances d_{hkl} of WSi_2 for PM target, cast target, and thin film deposited on Si(100) substrate by laser ablation and magnetron sputtering of cast target.

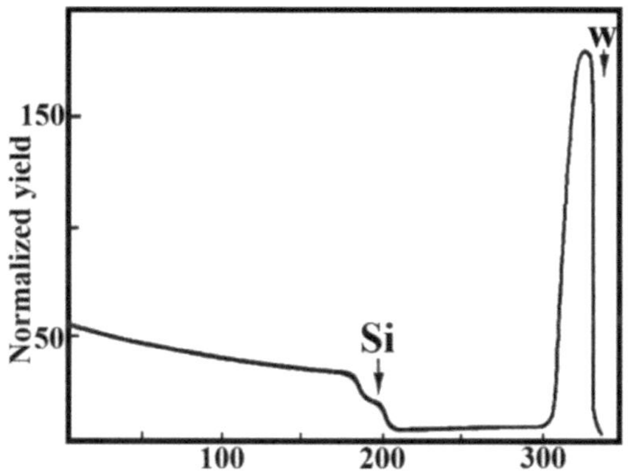

Figure 19. Rutherford backscattering spectra (E_0=1.5 MeV) for film structure WSi_2/Si. Arrows indicate energy positions corresponding to occurrence of W and Si atoms on the surface.

that a boundary between WSi_2 film of 100 nm thick and Si substrate is rather distinct. Similar experimental results are obtained when WSi_2 films are deposited by magnetron sputtering of cast composite targets. Auger electron spectrometry studies of WSi_2 films have revealed a sharp decrease in O and C contents during ion etching of specimens and deepening an analyzing zone. The behavior of O and C at WSi_2 layer is similar. W and Si are homogeneously distributed in the film, which is probably a consequence of interdiffusion. Near SiO_2/WSi_2 interface, there is a rather narrow layer where O is sharply increased and W is sharply decreased; however, Si content after a short decrease remains at a former level. C content remains at the same level as in WSi_2 layer and at the interface. This can be related to the limited solubility of C and C diffusion coefficient in initial WSi_2 film, transition layer, and SiO_2 sublayer. There are no visible changes of these layers after chemical etching as well as no changes of them after etching off the BPSG layers. However, after chemical etching and reactive ion etching, there is a light erosion of the WSi_2 surface. It is found that about 15% of the layer is etched off and the etched surface has a light roughness. The BPSG layer under the WSi_2 layer is partially melted during annealing at 1000°C, so the WSi_2 conducting paths are slightly uneven (curved); however, there are no fractures or cracks. The samples annealed at 900°C have no changes at all. The most visible changes in contacts to the poly-Si and wafers of the samples annealed at 900°C show the appearance of clear distortions in contact sites of wafers without the poly-Si sublayers. The wafers with poly-Si sublayers have no changes. This effect is greater on the samples annealed at 1000°C.

The mean electrical resistance of WSi_2 conducting paths before annealing is 33 kΩ. The mean electric resistivities of WSi_2 conducting paths with poly-Si sublayers are 6.32 and 4.7 Ω/□ after annealing at 900 and 1000°C, respectively. The mean electric resistivities of WSi_2 conducting paths without poly-Si sublayers are 6.65 and 5.6 Ω/□ after annealing at 900 and 1000°C, respectively. Electrical contacts to *p*-type layers are absent, because the layers in the vicinity of contacts can be heavily doped with P from the BPSG layer through the silicide layer or poly-Si. The electrical resistance of contacts to *n*-type conductivity layers on samples without poly-Si sublayers under WSi_2 layers is not reproducible and is very scattered with values up to 20 μΩ cm for 50 contacts. The samples with poly-Si sublayers exhibit much less scatter in a specific resistance. Thus, R_{Al-n+} = 360–400 kΩ for 50 contacts on a sample annealed at 900°C; however, the resistance for 50 contacts on a sample annealed at 1000°C is 1.4–1.8 kΩ. $R_{Al-polySi}$ = 600–700 kΩ for 100 contacts on samples annealed at 900°C; however, the analogous value for 100 contacts on samples annealed at 1000°C is 3.3–3.5 kΩ. All n^+-p-junctions on the large perimeter area of about 16,000 μm^2 and 2120 contacts have leakage currents lower than 0.04 nA at a bias voltage of 15 V. With an increase of the bias voltage, leakage currents increase monotonically up to 0.1 nA at 20 V; however, at 21–22 V junctions have breakdowns.

Analysis of experimental results suggests that WSi_2 deposited during laser evaporation do not form due to Si diffusion from the substrate. This view is supported by experiments on a film deposition on neutral substrates (MgO, Al_2O_3, ZrO_2) where the dominant phase is WSi_2. An important advantage of laser evaporation and magnetron sputtering for depositing films, compared to the annealing technology, is that single-phase films are deposited at a substrate temperature of 700°C, which is 300°C lower than in the case of annealing. Noteworthy is the fact that the tetragonal phase of WSi_2 is obtained at a substrate temperature of 600°C, which is in disagreement with data, which state that this phase nucleates and grows at temperatures above 620°C. The results on room-temperature deposition of thin (30 nm) films on Si substrates (and subsequently annealing at 750°C for 1 min) are also of interest. Their phase composition is analogous to that of films deposited directly on hot substrates. This is very promising technologically. Data on Rutherford backscattering and layer-by-layer X-ray phase analysis suggest that no transition layer exists between the film and the substrate. This is also an advantage compared to the annealing technology. Tetragonal disilicide films deposited by laser evaporation have the resistivity of 50–70 μΩ cm, which conform to values for films deposited by annealing, but it is much lower than for films deposited by magnetron sputtering of PM targets (120 μΩ cm). W and Si are homogeneously distributed across the film, which is probably a consequence of their interdiffusion. Near WSi_2/SiO_2 interface, there is a rather narrow transitional layer, where O and W replace each other. In SiO_2 layer, following the interface, mainly Si and O are found, with low amounts of W and C. Both of these elements are distributed homogeneously in SiO_2 layer by diffusion. O content in WSi_2 layer is a consequence of the fast O diffusion, compared with a self-diffusion of other components of the system, and the high O solubility in W as well as in its silicides.

2.7. Depositing titanium films

Of particular interest are thin films of $TiSi_2$ which due to their low electrical resistivity are the most promising for the inter-element wiring and silicidation areas of the emitters, collectors, and injectors of the bipolar IC, as well the sources, drains, and gates in VLSI. In this connection it is necessary to determine the conditions of formation of $TiSi_2$ films on substrates of single- and polycrystalline Si varying degrees of doping. Here, studies are done of thin films of the $TiSi_2$ formed by a solid-phase reaction of Ti films with the mono- and polycrystalline Si substrate. Thin films of $TiSi_2$ are obtained by magnetron sputtering targets of highly pure Ti, produced by multiple EB vacuum-melting Ti rods and followed by a heat treatment of the film obtained under nitrogen with O content of 10^{-4}%. The heat treatment is conducted in a diffusion furnace. To reduce the amount of O in the as-deposited Ti films, an annealing is used together with getters. Getters are Si wafers; the front and rear sides of which are coated with Ti films of a thickness 0.1 μm. Plates are positioned before and after experimental samples relative to Ar flow. As substrates, single-crystalline wafers of *n*-Si(100) of 100 mm in diameter, doped with P to the resistance of 4.5 Ω cm, are used. In the case of Ti silicides on the polycrystalline Si, Ti films are deposited on the oxidized beforehand poly-Si of 0.45 μm thick obtained by a silane ammonolysis under a reduced pressure. The thickness of Ti films is measured by the interference method with accuracy of 0.005 μm. A silicide film thickness is calculated from known ratios, knowing the thickness of the metal film, and is measured in a scanning electron microscope of the vertical cleavage of $TiSi_2$/Si structure. The Si substrate temperature during depositing Ti films does not exceed 423°C. Plates are subjected to the standard chemical treatment, and immediately prior to Ti deposition, Si surface is purified by plasma etching. To investigate the degree of influence of both the alloying and impurity-type doping, Si substrates (or a poly-Si film) are subjected to ion doping with B or P at ion energies of 40 and 60 keV, respectively. An annealing Ti film to form $TiSi_2$ is carried out in a diffusion furnace in two steps: at 898K in a nitrogen atmosphere for 30 min and, then, after removing the TiN layer, at 1123K and 30–60 min in a vacuum or Ar atmosphere. A need to anneal in nitrogen causes by the fact that the resulting surface layer consisting essentially of TiN prevented the lateral Ti diffusion which in the real VLSI structures can cause short circuits. The crystal structure of the elemental and phase composition of as-deposited films and silicide films are studied by electron diffraction, scanning and electron microscopy, Rutherford backscattering, as well as X-ray methods, in particular by "moving beam," which allows determining the change of the phase composition across the film thickness. A study is performed on the profiles of the element distribution across the film thickness by Rutherford backscattering with an initial energy of 1 MeV He ions. The scattering angle is 170°. The energy resolution of the detector system is 17 keV. To improve the depth resolution, a sliding experimental geometry is used, in which angles of incidence and scattered ions have a value of 60°. This improvement allows reaching a resolution of 10 nm for Ti and 15 nm for Si, while at normal resolution geometry, they are 20 and 30 nm, respectively. **Figure 20** shows energy spectra of Rutherford backscattering for as-deposited and annealed at 898 K samples. It can be seen (**Figure 20**, spectrum 1) that in the original Ti film, some O is present, as evidenced by the spectrum peak in 350–368 keV energy. The profiles of an element distribution, calculated from the spectrum, allow the thickness of the film to be determined, which is found to be ~3×10^{17} at./cm^2 for the sample. Provided that the volume Ti content is 5.68×10^{22} at./cm^3 and $d = x$ (at./cm^2)/N^* (at./cm^3), Ti film thickness is found to be ~53 μm. O mainly localized in the surface layer of the film of a thickness 1×10^{17} at./cm^2 (15–20 nm), and its concentration does not exceed 25%. An annealing under nitrogen at 898 K and 30 min leads to a redistribution of elements in the surface layer (**Figure 20**, spectrum 2). A layer near the surface is enriched with N atoms and with a small amount of O. At a range of depths (3.5–5.5)$\times10^{17}$ at./cm^2 (a conversion to the unit of length in this case is not correct due to a lack of accurate data on the volume concentration of atoms in the silicide film), the relative concentration of elements corresponds to $TiSi_2$ phase. After chemical or plasma-chemical removal of the fluorine-containing plasma, the surface layer of the film has a phase composition of the $TiSi_2$ with some excess of Si. In

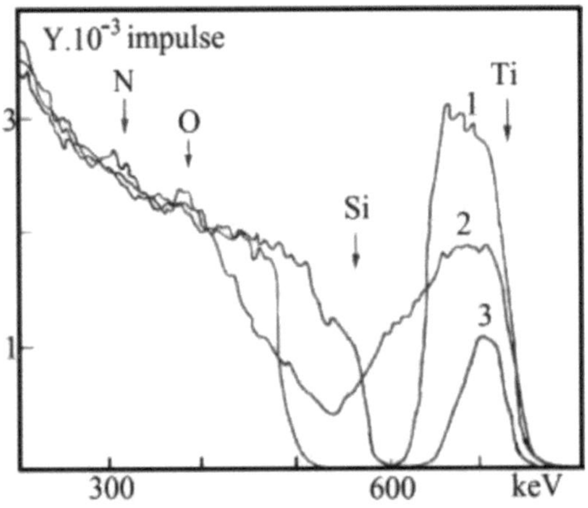

Figure 20. Energy-loss RBS spectra for as-deposited and annealed at 898 K samples. Ti (1) and $TiSi_2$ (2, 3), 2, 3—898K; 2—15 min; 3—15 min after etching off upper layer.

this case, the films do not contain N and O in amounts sufficient for detection by RBS. As a result of X-ray studies, it is established that the formed $TiSi_2$ phase has an orthorhombic C54 structure with lattice constants a = 0.825 nm, b = 0.4783 nm, and c = 0.854 nm, wherein Ti_5O_9 phase is detected on the surface of the film (α = 0.5, 1 deg). The phase composition of films is enough uniform in thickness.

There is no significant difference for $TiSi_2$ films formed on n- and p-type substrates. After the plasma-chemical removal of the surface layer, $TiSi_2$ film is amorphous, and after re-annealing at 1123K, is polycrystalline with an average grain size 0.1–0.3 μm. After the removal of $TiSi_2$ film, precipitates are found on the substrate surface which can be easily removed mechanically. An analysis of these precipitates by electron diffraction has revealed that it is TiN, resulting in the surface film due to its annealing in a nitrogen environment. These results are consistent with RBS (**Figure 20**, spectrum 2). Studies, as well as microreliefs of the substrate surface after the removal $TiSi_2$ film, lead to the conclusion about a fairly clear boundary between $TiSi_2$ film and Si substrate. It can be explained by the mechanism of an interaction of Ti with Si, at which a diffusing dopant is predominantly Si. A second stage of annealing the film at 1023K during 30 and 60 min, after the removal of TiN surface layer, leads to a stabilization of C54 $TiSi_2$ phase. Under these conditions, $TiSi_2$ films have a homogeneous phase composition across their thickness and did not contain the phase of Ti_5O_9. Seemingly it had been removed together with the surface layer. The resistivity of the films of C54 $TiSi_2$ is 13–15 μΩ cm, which is in agreement with known data. With increasing of the substrate doping with P or B, any changes in the phase composition of films are not established: only forming the low-impedance phase C54 $TiSi_2$. It is shown that the low-impedance phase C54 is formed at 963 K during 10–30 min, and its formation is dependent on O content in as-deposited Ti films. Reducing the temperature of the formation of the phase C54 in $TiSi_2$ to 898 K, in this case, may be linked with a sufficiently low O content in as-deposited Ti films. This in turn is determined by the method of producing Ti targets and characteristics of the heat treatment of films associated with additional getters.

2.8. Studying $TiSi_2$/Si interfaces

2.8.1. *Short background*

A study on the interaction of Ti atoms with a Si surface as chemical reactions at the interface between thin films of Ti and Si is of great interest. It is known that at Ti-Si interface at above 700°C, only $TiSi_2$ can be formed. It is also proved that on an amorphous Si substrate, the growth rate is proportional to the square root of the annealing time (so-called $t^{0.5}$ rule). On a single-crystalline Si substrate, the growth rate of $TiSi_2$ is much slower, and grown $TiSi_2$ films are uneven in a thickness if Ti-Si interface contaminated with impurities. Annealing Ti films on the single-crystalline Si in oxygen or in a mixture of oxygen + nitrogen at 600°C leads to a formation of TiO_2 and $TiSi_2$; however, a heating in N_2 leads to the formation of TiN, Ti_5Si_3, and TiSi without $TiSi_2$. Whereas if to anneal a mixture of Ti_5Si_3 and TiSi at >800°C, it transforms to $TiSi_2$. It is also interesting that at room temperature, Ti and Si do not interact at Ti-Si(100) interface. From electron spectroscopy, similar data for Ti-Si(111) system are also known. However, by RBS experiments a diffuseness of Ti-Si(111) interface at room temperature has been discussed. The main reason of this obstacle for solving the problem is a lack of suitable instruments to identify the reaction products in a Ti-Si system. These instruments should be similar in their capabilities, especially in the depth of the layer to be analyzed. They should be based on a study of Si($L_{2,3}$VV) Auger line shape of, e.g., PtSi system. Thus, it seems that the most distinguished technique which can help us in eliminating the problem is an electron energy-loss spectroscopy (EELS) which is known by its high sensitivity to electronic restructuring of solid surfaces and by possibilities to vary a depth of the layer under analysis [28]. The goal of this study is to obtain electron energy-loss spectra of $TiSi_2$, Ti, and Si and to estimate analytic possibilities of EELS in a study of chemical transformations at Ti-Si interface.

2.8.2. *$TiSi_2$/Si interfaces*

Single-crystal Si(100) wafers doped with P and Ti rods after EB floating zone melting are used to prepare $TiSi_2$ films by a standard self-aligned silicide method in Ti-Si(100) system. Si wafers with the electrical resistance of 1 Ω cm are previously chemically cleaned from organic impurities and then immersed in HF, followed by washing in a deionized water to remove a natural surface SiO_2. A Si substrate temperature during Ti deposition is about 180°C. The thickness of Ti films is 40–80 nm. $TiSi_2$ films are deposited by magnetron sputtering of Ti targets. A subsequent silicide formation is done by the two-stage high-temperature annealing. The first annealing is carried out at 600–700°C under a N_2 atmosphere for 35–40 min. An excess unreacted Ti, TiO_2, and TiN are removed using an etchant selective to silicide. Silicidation and homogenization are completed at a second annealing at 800°C under Ar atmosphere. The thickness of $TiSi_2$ layers is about 100 nm. The sheet resistance of the films is measured by a four-point probe. The typical resistivity of the Ti layers is 50 μΩ cm; the resistivity of the $TiSi_2$ layers is 13 μΩ cm. An analysis of trace impurities is performed by high sensitive physical methods, such as fast neutron activation, deuteron activation, He^3 ion activation, mass spectrometry with inductively coupled plasma, etc. When registering electron spectra in the analog mode, the relative energy resolution is 0.6%; in a retarding mode, the absolute resolution does not depend on the energy and is of 0.7 eV. Electron spectra are measured in a spectrometer with a double-pass cylindrical mirror analyzer. To get Auger electron spectra, a monoenergetic (half width, 0.5 eV) electron beam is used with an energy of E_p = 3 keV. For electron energy-loss spectra (EELS), electron beams with energies 100, 400, and 2000 eV are used. The removal of surface contaminations and analysis of the depth distribution of elements are conducted in combination with sputtering of the sample surface by 2 keV Ar ions with 9 μA cm^2. A residual gas pressure in the test chamber of the spectrometer is $5{\times}10^{-8}$ Pa, and Ar pressure during sputtering was $3{\times}10^{-3}$ Pa. An elemental analysis is performed by Auger spectroscopy using factors of an elemental sensitivity.

The electrical properties of metal and silicide layers in Ti-Si system depend strongly on impurities. Thus, it is very important to use Ti targets with the highest purity for the sputter. This means a low content of gas-forming interstitials and metal impurities. After a complex of vacuum procedures, Ti designed for magnetron sputtering contains trace impurities according with **Table 1**. In **Figure 21** a distribution of elements in $TiSi_2$ films is shown. A subsurface area, after sputter over 9 min, contains, besides $TiSi_2$, certain amounts of TiN and SiO_2. These data can be drawn from an analysis of the form Si ($L_{2.3}$VV) lines and an intensity ratio of Ti (LMV) (418 eV) to Ti (LMM) (387 eV). A composition of the interior of films, sputtered for 30 min (from 10th min etching up to the 40th min), is close to $TiSi_2$. A total content of gas-forming interstitials (C, O) in this area of films is not greater than 1–2 at.%. Note that the ratio of peak intensities of Ti (418 eV)/Ti (387 eV) lines in $TiSi_2$, measured at a modulation voltage of 3 V, is equal to 1.13. This value is very close to 1.2. For metal Ti this value is 1.6; thus, an intensity ratio of Ti (418 eV)/Ti(387 eV) may be used for analytical purposes when estimating a number of Ti atoms in reaction with Si. Si($L_{2.3}$VV) Auger line shape in the transition region from $TiSi_2$ to Si varies insignificantly at layer-to-layer sputtering. Such behavior is known as a characteristic of refractory metals silicides [29, 30]. It is noted that in the range of kinetic energies from 75 to 85 eV, however, variations of electron spectra nevertheless happened. It seems that this fact is difficult to use for analytical purposes because of a low relative intensity of this spectral region. Thus, a shift of Si ($L_{2.3}$VV) line by 1.3 eV, observed in the initial stage of the formation of $TiSi_2$, seems to be due to the fact that relaxation conditions during the Auger process in

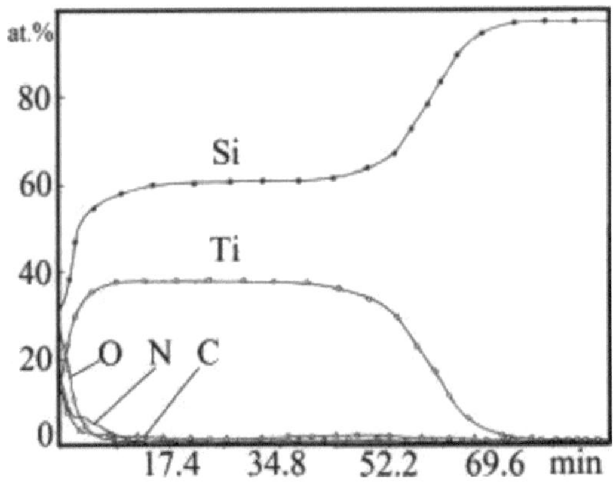

Figure 21. Profile element distribution in $TiSi_2$/Si(100) structure, obtained by AES during etching with argon ions.

the thin Si layer on the surface of Ti or $TiSi_x$ films differ from those in a bulk Si. The changes observed in the low-energy part of the Auger spectrum are due to changes in Auger energy-losses spectra. Therefore, EEL spectra are studied using a monochromatic electron beam as an excitation source. EEL spectra, which are obtained in reflection geometry, provided data on excitations in layers of 0.3–2 nm thick. Thus, a contribution of surface excitations, as a rule, is of a great importance for these spectra. The value of this contribution depends on both the mean-free path of electrons in analysis and the experimental geometry. In addition to a quantitative interpretation of reflection, EEL spectra should be taken into account not only multiple small-angle electron scattering but large-angle scattering as well.

For analytical purposes, some of the parameters of the measured spectra are often possible to use without further mathematical processing. For example, energy positions of major peaks in EEL spectra are slightly depended on a shape of multiple scattering histories. It is of interest to compare measured energy locations of intense peaks in EEL spectra of $TiSi_2$ at various energies of primary electron energies of bulk and surface plasmons, estimated in a free-electron approximation: $\omega_p = (4\pi ne^2/m)^{1/2}$, where n a density of valence electrons per volume unit, $\omega_s = \omega_p/\sqrt{2}$ (see **Table 6**). An estimation of the valence electron density is carried out using known values of specific densities $\varrho_{Ti} = 4.51$ g/cm³ and $\varrho_{si} = 2.33$ g/cm³. For $TiSi_2$ a density of 4.043 g/cm³ measured by X-ray analysis is used [25]. It should be emphasized that tabulated energy values differ from maxima locations in EEL spectra recorded in a form of the second derivative $d^2N(E)/dE^2$ for $E_p = 153$ eV [31]. A comparison of energy shows that a location of the main peak in EEL spectrum with $E_p = 100$ eV of $TiSi_2$ is very close to the surface plasmon energy $\hbar\omega_s = 13.9$ eV and with $E_p = 2$ keV, which is close to the bulk plasmon energy. The contributions of both the bulk and surface plasmons in EEL spectrum at $E_p = 400$ eV are comparable in a magnitude. As the energy of primary electrons increases, spectral peaks become narrower. A total width at a half height of the mean peak, $\Delta_{1/2}$, received by this method, is decreased from 19 eV at $E_p = 100$ eV to 10 eV at $E_p = 2$ keV. This parameter is influenced by the background of electron inelastic scattering. Thus, a peak width at 3/4 height, $\Delta_{3/4}$, seems more analytical. A comparison of EEL spectra of Ti and Si with the spectrum of $TiSi_2$ for different energies of primary electrons is done. The spectra differences can be observed for all E_p, but the most analytic spectra are spectra with narrow peaks.

Considering spectra at $E_p = 100$ eV, when a mean-free electron path is minimal—0.4 nm, the locations of the main maxima in the spectra are different, but due to a complicated shape of spectra, the energy locations cannot be the main analytical functions. It seems that more characteristics are widths $\Delta_{3/4}$ that are equal to 16.3, 13.7, and 10.5 eV for spectra of Ti, $TiSi_2$,

Sample	M			$\hbar\omega_s$	
	100	**400**	**2000**	**Ref. [31]**	$(4\pi ne^2/m)^{1/2}$
Si	15.2	16.8	17.0	16.8	16.6
Ti	8.4	14.6	17.4	19.2	17.7
$TiSi_2$	13.2	18.0	19.0	17.2	19.8

Table 6. The maximum positions M of the EEL spectra and plasmon energies $\hbar\omega_p$ (in eV).

and Si, respectively, at E_p=100 eV. By measuring values for thin Ti-Si films, their presence can be found in the layer of 0.4 nm thick. Of course, there is a problem of distinguishing a simple mixture of Ti+Si and $TiSi_2$ with EEL spectra. The artificial spectrum of a simple unreacted Ti+Si mixture is produced by a superposition of EEL spectra normalized to the elastic peak intensity and taken with corresponding weights. It can be shown that at low primary electron energy (E_p=100 eV), when a peak width is greater, a value of $\Delta_{3/4}$ close to that for $TiSi_2$ may be received at a certain mixture composition. In this case, the simple unreacted Ti+Si mixture and a composition of $TiSi_2$ layer after a chemical reaction may be distinguished, because (*a*) the location of the maximum in the spectrum of the mixture is close either to the location of the maximum in Ti spectrum (8.4 eV) or Si spectrum (15.2 eV), excluding a small transition region of compositions near Ti-Si=1; (*b*) with coincident maxima in spectra, the *a* location on the energy-loss scale corresponds to 29.9 eV for $TiSi_2$ and 24.6 eV for the mixture of Ti+Si spectrum. With E_p = 2 keV, it is not difficult to distinguish EEL spectrum of $TiSi_2$ from those of the mechanical Ti+Si mixture, because for mixture spectra the location of the main maximum are within 17–17.4 eV, whereas for $TiSi_2$ this value is 18.8 eV (**Figure 22**). Besides, the EEL spectrum of $TiSi_2$ exhibits also a second maximum at energy of ≈38 eV that is missing in spectra of Ti-Si mixture. Thus, EELS enables one to carry out an analysis of an interaction of components in Ti-Si system with a high depth resolution.

2.9. Depositing cobalt disilicide films by laser ablation

2.9.1. Short background

An interest in silicides of refractory metals has increased significantly due to their great potential as a material of low resistance contacts, gates, and interconnects of the thin-film metallization of Si integrated circuits. The principal possibility of obtaining pure $CoSi_2$ targets by vacuum-metallurgical methods is developed [15, 32]. The optimal conditions of sputter targets have been revealed, providing a deposition of $CoSi_2$ films with the specific resistivity of 30 μΩ cm. By X-ray

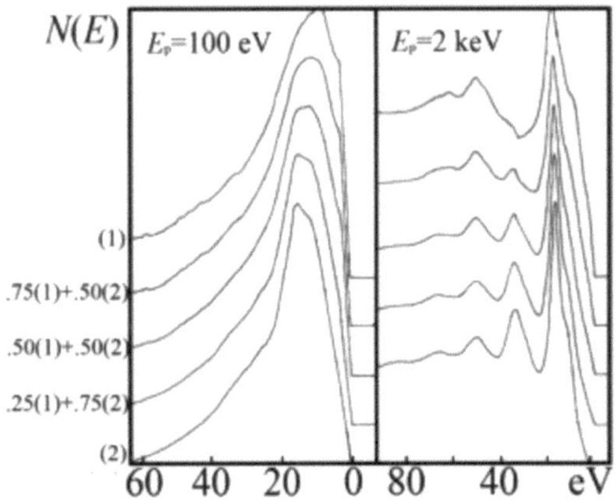

Figure 22. Electron energy-loss spectra of Ti (1) and Si (2); a linear combination of these spectra with different weight fractions.

diffraction patterns in the sliding beam incidence and Rutherford backscattering of He ions, the phase and elemental composition of the films are investigated. Currently in microelectronics two methods are used for producing silicide thin films. The first of them is a deposition of metal films on a Si surface, followed by high-temperature annealing. The second is composed of sputtering targets made of a silicide powder. Unfortunately, the manufacture of targets using PM technologies usually leads to instabilities of sputtering and contamination of targets, which increase the specific resistivity of deposited films. Thus, vacuum-melting procedures, seemingly, become an optimal and logic decision for the production of targets. In recent years, interesting results have been obtained by the method of sputtering metal and semiconductor films by a laser radiation which has several advantages over other methods of depositing films. It is compatible with high vacuum setups and does not require using a working gas. When the laser radiation flux density in the affected area exceeds 10^{-9} W s^{-2}, the evaporation of the target material takes place without a formation of a liquid phase (laser ablation), so that the stoichiometry of the film corresponds to the composition of the target. This is an excellent opportunity of using laser deposition films for substances with a complex stoichiometry. A high efficient sputter rate (10^{-6} A s^{-1}) supports reducing the contamination of the film by the residual gasses of the vacuum system. Due to the high energy of condensed atoms (10^5 K), it is possible to obtain epitaxial films at lower substrate temperatures than for other methods. Here, experimental results of our studies are presented of obtaining the $CoSi_2$ films by laser ablation of cast high-purity targets.

2.9.2. Depositing cobalt disilicide films

As the initial materials for preparing $CoSi_2$ targets are used a polycrystalline Si of 99.999% purity and a commercial Co. To further increase the purity of Co, a double EB vacuum refining is performed [15]. A typical content of impurities after chemical and EB floating zone purifying is shown in **Table 1**. For the preparation of $CoSi_2$ of a stoichiometric composition, the mixture of high-purity Co and Si is melted in a vacuum induction furnace in high purity Ar. The melt is poured into molds receiving rods of 10 mm in diameter and 150 mm in length. The resulting bars are subjected to double floating EB zone melting. $CoSi_2$ targets are of 20×15×5 mm. They are evaporated by laser in a vacuum setup. A solid-state Nd-glass laser is used with wavelength of 1.06 μm, pulse repetition rate of 10 Hz, and pulse energy of 0.8 J. The vacuum chamber is evacuated to a vacuum of 1×10^{-6} Torr using a turbo pump. The laser beam is focused by a lens; the focal spot size is 2–3 mm. A distance between the target and substrate is 70 mm. For heating the substrate, an infrared heater with halogen lamps is used. The Si(100) wafers, mica, MgO, Al_2O_3, and ZrO_2 are used as substrates. Substrates are subjected to chemical cleaning and then immediately prior to depositing are heated at 850° for 10 min in vacuum. The temperature of substrates is varied in the range of 100–800°; the deposition time is varied from 5 to 20 min. A film thickness is measured using a profilometer and is from 300 to 1500 Å. The specific resistivity of bulk samples and thin films is measured by the four-point probe method. XRD data of $CoSi_2$ samples show that they are a single phase and composed of stoichiometric $CoSi_2$. A local X-ray analysis of cross sections has revealed the homogeneity of the chemical composition of $CoSi_2$ samples. Double EB melting of $CoSi_2$ ingots does not lead to a significant change in its phase composition. The specific resistivity of bulk samples of $CoSi_2$ is 16–20 μΩ cm, which is close to the data, obtained for single crystals of $CoSi_2$. The samples prepared by the PM technology have an electrical resistivity of 65–68 μΩ cm. An important advantage of the method of depositing films by the laser ablation compared with annealing a Co-Si system is a deposition of films at a substrate temperature of 700–800°C. This temperature is 200–300°C lower than in the case of annealing. In addition, the film does not contain Ar, which usually occurs in magnetron sputtering. The results of XRD of the cast target and films obtained in the range of the substrate temperature from 200 to 750°C show that as-deposited films are textured, and with increasing the substrate temperature, the degree of texturing increases (when the temperature increases, the amount of equiaxed polycrystalline grains decreases sharply). At the substrate temperature of 400°C, films are disilicides $CoSi_2$ (**Table 7**). The films deposited at other substrate temperatures are bi-phasic, and at 200°C together with the $CoSi_2$, which is the main phase, an excess of Si is presented as well. At temperatures of 600, 700, and 750°C together with the main phase ($CoSi_2$), a small amount of CoSi is found. Sputtering onto substrates ZrO or MgO at 700°C also affords films consisting of $CoSi_2$ and a small amount of CoSi. From RBS, at the films deposited at the substrate temperature of 600°C, the ratio of Si/Co is about 2 (**Figure 23**). The

hkl	$CoSi_2$		Target		Film	
	d_{tab}	*I*	*d*	*I*	*d*	*I*
111	3.096	90	3.099	80	3.118	82
220	1.898	100	1.897	100	1.897	82
311	1.618	23	1.616	24	1.620	14
222	1.548	1	–	–	–	–
400	1.340	17	1.340	12	1.346	13
311	1.230	11	1.230	10	1.231	18
422	1.095	2	1.094	20	1.094	22

Table 7. Interplanar distances of $CoSi_2$ for the cast target and the film deposited on the Si(100) substrate at 400°C.

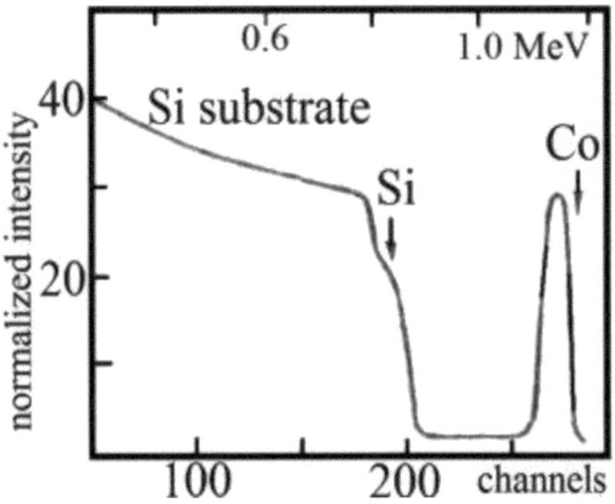

Figure 23. RBS spectra of the $CoSi_2$ film structure. Arrows indicate the position of energy corresponding to surface bedding Co and Si atoms.

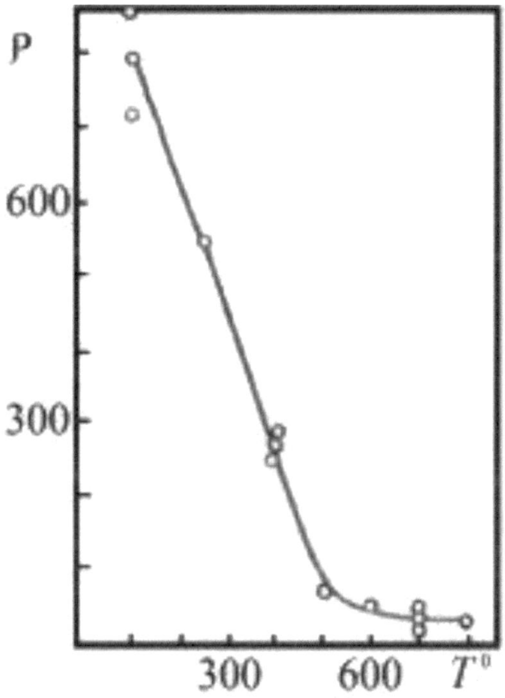

Figure 24. The dependence of the specific resistivity of $CoSi_2$ films on Si(111) on substrate temperature.

thickness of the stoichiometric film is about 600 Å. The transition layer between the film and the substrate is not detected. **Figure 24** shows the dependence on the substrate temperature of the specific resistance of $CoSi_2$ films of 600 Å thick on Si(100). It is seen that with increasing the substrate temperature, the specific resistivity of the film decreases. Especially sharply it is at the range of 100–500°C where the specific resistivity is reduced by about one order of magnitude (from 800 to 60 μΩ cm). In the range of 600–800°C, the specific resistivity depends weakly on the temperature and reaches the level of 30 μΩ cm. The relatively high specific resistivity values of $CoSi_2$ films at low temperatures can be attributed to the presence of the excess Si in the composition of films. At high temperatures, the predominance of the substrate composed of $CoSi_2$ films, as well as increasing the degree of texture of samples, leads to a sharp decrease in their resistivity, which explains the behavior of the curve depending on the specific resistivity.

3. Key findings

1. The multiple EB floating zone melting, multiple EB melting, vacuum levitation melting, and electric arc vacuum melting have been successfully used together with chemical techniques of a preliminary purifying by halides and ion exchange of refractory metals. Preparing highly pure refractory metals containing trace contents (on the level of ppm and ppb) of gas-forming interstitials as well as radioactive impurities and light metals

is demonstrated to be possible. To extend a range of the analyzing limitations, the advanced analytic techniques are used, which allowed to determine the real trace contents even on the lowest level—of ppb. It is also shown that both the production methods of cast targets and sputter conditions have a strong effect on the physical properties of deposited refractory metal thin films. By indicating ways to evaluate interstitials and other impurities in as-deposited films, this approach allows one to determine regimes of sputtering providing the deposition of high-purity metallic layers with the optimal specific resistivity.

2. Impurity contents in thin TiW films deposited by co-sputtering of cast metal targets are quite low comparing with impurity contents of films deposited by magnetron sputtering of PM targets under similar conditions. The specific resistivity of TiW thin films strongly depends on Ti/W ratio. The resistivity of thin films of pure W and Ti is close to tabulated ones for bulk metals. The resistivity of films increases gradually as the Ti/W ratio changes from those for the pure W to the pure Ti; however, it is worth to mention that the resistivity of quasi-alloy films is always higher than those of pure metals.

3. TiW films, deposited by magnetron sputtering of highly pure composite cast Ti/W targets, are grown and studied by AES, XRA, and XRD. AES data showed that Ti/W ratio in TiW films is nearly constant during the sputtering lifetime of composite targets. X-ray spectral analysis of contacts with TiW barrier layers, obtained at the increased sputtering power, is partly due to the higher dispersion of the microstructure. XRD data show that TiW films are solid solutions of Ti in W matrix with increased W lattice parameters. The structure is found to be a solid solution of α-Ti in *bcc* α-W with a lattice parameter of 0.318–0.323 nm. It is also shown that a relative increase in W lattice parameters depends on physical conditions of Si substrates or sublayers on which films are deposited.

4. Schottky diodes with TiW films exhibit a higher level of current leakage than diodes with Mo films, although in both cases, leakage currents are not an obstacle for the commercial use of these highly pure refractory metals as Schottky diode barrier layers. The great advantage of Schottky diodes with Si/PtSi/TiW/Al structure is a high thermal stability of the direct breakdown potential.

5. The refractory metal silicides are very brittle. Thus, one of the experimental approaches to produce these silicides would be used to cast refractory metal silicide targets by HF levitation melting and attaching cast silicide pieces to copper bases by ultrasonic soldering. Vacuum melting (and casting) of silicides solves three main problems of the silicide thin-film deposition: easy production of samples of desired geometries, any chemical compositions, and high purity of silicide thin films. For magnetron sputtering, the composite cast WSi_2 and $MoSi_2$ targets of 152 mm in diameter are produced. Our study has revealed perspective possibilities of depositing WSi_2 films by both the laser evaporation of small cast targets and magnetron sputtering of composite cast WSi_2 targets. The conditions and regimes of the laser evaporation and magnetron sputtering of cast targets ensure the formation of single-phase WSi_2 films with the optimal specific resistivity of 50–70 $\mu\Omega$ cm. Films about 200–250 nm thick are shown to be effective diffusion barriers and conducting paths at annealing temperatures up to 900–1000°C. Leakage currents do not deteriorate diode junctions having WSi_2 layers with poly-Si sublayers.

6. The principal possibilities of obtaining high-purity cast targets of $CoSi_2$ and WSi_2 by crystallization from its liquid phase using a set of metallurgical methods are studied. The modes of the laser evaporating (ablation) have revealed to ensure a stable deposition of stoichiometric $CoSi_2$ and WSi_2 films with the specific resistivity of 30 and 50 μΩ cm, respectively. The results of XRD of cast targets and films obtained in the range of the substrate temperature from 200 to 750°C show that the as-deposited films are textured and with increasing substrate temperature, the degree of texturing increases (when the temperature increases, the amount of equiaxed polycrystalline grains decreases sharply). It should be emphasized that at a substrate temperature of 400°C, as-deposited films are stoichiometric disilicides $CoSi_2$ and WSi_2.

Author details

Vadim Glebovsky

Address all correspondence to:glebovs@issp.ac.ru

Institute of Solid State Physics, The Russian Academy of Sciences, Russia

References

[1] Benzing WC. Shrinking VLSI dimensions demand new interconnection materials. Electronics. 1982;**55**(17):116-119

[2] Ortner HM, Wilhartitz P, Grasserbauer M, Virag A, Friedbacher G. Ultrapurity in metallurgy—Facts and fiction. Kontakte (Darmstadt). 1987;**3**:3-7

[3] Ortner HM, Bloedorn W, Friedbacher G, Grasserbauer M, Krivan V, Virag A, Wilhartitz P, Wuensch G. Ultrapurity in metallurgy—With special reference to refractory metals application in microelectronics. Kontakte (Darmstadt). 1988;**3**:38-52

[4] Iwata S, Yamamoto N, Kobayashi N, Tereda T, Mizutani T. A new tungsten gate process for VLSI applications. IEEE Transactions on Electron Devices. 1984;**31**(9):1174-1179

[5] Yamamoto N, Kume H, Iwata S, Yagi K, Kobayashi N, Mori N, Miyazaki H. Fabrication of highly reliable gate MOS VLSI's. Journal of Electrochemical Society. 1986;**133**(2):401-407

[6] Hoffman V. Tungsten-titanium diffusion barrier metallization. Solid State Technology. 1983;**26**(6):119-126

[7] Sawada S. On advanced sputtering targets of refractory metals and their silicides for VLSI application. In: Proc. 12th International Plansee Seminar. Vol. 1; Reutte/Tirol, Austria, Austria. 1989. pp. 201-210

[8] Glebovsky VG. Preparation of high purity refractory metals for thin film metallization. In: Proceedings of International Symposium on Rare Metals. Vol. 1; Kokura, Japan. 1990. pp. 343-353

[9] Fromm E, Gebhardt E. Gase und Kohlenstoff in Metalle. Berlin: Springer Verlag; 1976

[10] Shipilevsky BM, Glebovsky VG. Competition of bulk and surface processes in the kinetics of hydrogen and nitrogen evolution from metals into vacuum. Surface Science. 1989;**216**:509-527

[11] Ishigami T, Ishihara H, Shimotori K. High purity Ti sputter targets for VLSI. Toshiba Review. 1987;**161**:38-41

[12] Klimenko GL, Blohin AA, Glebovsky VG, Ermolov SN, Mayorov DY, Kopiryn AA. Production of high purity W and Mo powders by ion exchange. Russian Metallurgy (Metalli). 2001;**3**:49-55

[13] Stinov ED, Sidorov NS, Glebovsky VG, Karandashev VK. The combined purification of titanium. Russian Metallurgy (Metalli). 2004;**6**:49-53

[14] Sidorov NS, Glebovsky VG, Shtinov ED. Refining of nickel by chemical & solidification methods. Russian Metallurgy (Metalli). 2011;**7**:610-615

[15] Glebovsky VG, Sidorov NS, Stinov ED, and Gnesin BA. Electron-beam floating zone growing of high purity cobalt crystals. Materials Letters. 1998;**36**:308-314

[16] Glebovsky VG. Physical and technological aspects of processing high-purity refractory metals. In: Glebovsky VG, editor. Recrystallization in Materials Processing. Rieka, Croatia: InTech Publishing House; 2015, 211p

[17] Amazawa T, Oikawa H, Shiono N, Honma N. Extended Abstracts of 16th Conference on Solid State Devices and Materials; Kobe, Japan. 1984; p. 269

[18] Yamamoto N. A study on the low resistivity gate electrode interconnections for ultra high density devices [doctoral thesis]. School of Engineering, Tokyo University; Tokyo University 1986

[19] Glebovsky VG, Markaryants EA. Thin film metallization by magnetron sputtering from highly pure molybdenum targets. Journal of Alloys and Compounds. 1993;**190**:157-160

[20] Glebovsky VG, Markaryants EA, Titov EV. Deposition of W-Ti thin films by magnetron sputtering. Materials Letters. 1994;**20**:89-93

[21] Babcock SE, Tu KN. Titanium-tungsten contacts to Si: The effects of alloying on schottky contact and silicide formation. Journal of Applied Physics. 1982;**53**(10):6898-6905

[22] Babcock SE, Tu KN. Titanium-tungsten contacts to Si: II. Its stability against aluminum penetration. Journal of Applied Physics. 1986;**59**:1599-1605

[23] Glebovsky VG, Yastschak VY, Baranov VV, Sackovich EL. Properties of tungsten-titanium thin films obtained by magnetron sputtering of composite targets. Thin Solid Films. 1995;**257**:1-6

[24] Glebovsky VG. Metallurgical aspects of preparation of high purity refractory metals for thin film metallization. In: Proceedings of 12th International Plansee Seminar. Vol. 3; Reutte/Tirol, Austria, Austria. 1989. pp. 379-389

[25] Murarka SP. Silicides for VLSI Applications. Academic Press, New York; 1983

[26] Glebovsky VG, Oganyan RA, Ermolov SN, Stinov ED, Kolosova EV. Preparation of tungsten disilicide thin films by laser evaporation. Thin Solid Films. 1994;**239**:192-195

[27] Glebovsky VG, Ermolov SN, Motuzenko VN, Stinov ED. Thin silicide films deposited from cast silicide targets. Materials Letters. 1998;**37**:44-48

[28] Shulga YM, Glebovsky VG, Dulinets YC, Rubtsov VI, Borodko YG. Electron energy loss spectroscopy as an analytical tool in the study of $TiSi_2$/Si interfaces. Materials Letters. 1993;**15**:325-330

[29] Butz R, Rubloff GW, Tan TY, Ho PS. Chemical and structural aspects of reaction at the Ti/Si interface. Physical Review B. 1984;**30**(15):5421

[30] Raaijmakers IJMM. Fundamental aspects of reactions of titanium-silicon thin films for integrated circuits [PhD thesis]. Philips Research Laboratories, Eindhoven; 1988

[31] Sharma JKN, Chakraborty BR, Shivaprasad SM. Chemical shifts of Si and Ti in $TiSi_2$ studied by AES, SEELS and IXPS. Journal of Vacuum Science & Technology A. 1988;**6**:3120

[32] Glebovsky VG, Oganyan RA, Ermolov SN, Kolosova EV. Deposition of cobalt disilicide thin films by laser ablation. Thin Solid Films. 1994;**248**:145-148

Section 4

Applications

Chapter 5

Operational Amplifier Design in CMOS at Low-Voltage for Sensor Input Front-End Circuits in VLSI Devices

Muhaned Zaidi, Ian Grout and Abu Khari A'ain

Additional information is available at the end of the chapter

http://dx.doi.org/10.5772/intechopen.68815

Abstract

Today, digital circuit cores provide the main circuit implementation approach for integrated circuit (IC) functions in very-large-scale integration (VLSI) circuits and systems. Typical functions include sensor signal input, data storage, digital signal processing (DSP) operations, system control and communications. Despite the fact that a large portion of the circuitry may be developed and implemented using digital logic techniques, there is still a need for high performance analogue circuits such as amplifiers and filters that provide signal conditioning functionality prior to sampling into the digital domain using an analogue-to-digital converter (ADC) for analogue sensor signals. The demands on the design require a multitude of requirements to be taken into account. In this chapter, the design of the operational amplifier (op-amp) is discussed as an important circuit within the front-end circuitry of a mixed-signal IC. The discussion will focus on the design of the op-amp using different compensation schemes incorporating negative Miller compensation and designed to operate at lower power supply voltage levels. A design case study is included which utilises the g_m/I_D ratio design approach to determine the transistor sizes. The simulation approach is focussed on the open-loop frequency response performance of the op-amp.

Keywords: op-amp, design, stability, g_m/I_D, Miller compensation, negative Miller compensation

1. Introduction

In this chapter, the focus of the discussion is on the design of the op-amp, which will act as an integral part of the on-chip analogue signal conditioning circuitry for the front-end section of a mixed-signal IC. The performance requirements and design issues for circuit operation on a single-rail power supply and operating at 3.3 V or lower will be considered. The op-amp architecture will be discussed, and the focus will be on the design of the compensation circuitry that will be required for amplifier stability purposes. In particular, the use of Miller and negative Miller compensation techniques, and the effects of different compensation techniques on amplifier operation, will be identified. The discussion will be supported using suitable simulation study results. The chapter will initially consider the analogue circuit requirements before discussing op-amp design and compensation techniques. The concepts introduced and analysed will be accompanied by analogue circuit simulation results using Cadence Spectre simulator and the circuit design will be implemented using a 0.35 μm n-well complementary metal oxide semiconductor (CMOS) fabrication process. In order to provide a better understanding, the discussion will include the use of MATLAB for mathematical modelling the frequency response of the op-amp in open loop.

2. Analogue front end circuits in mixed-signal IC designs

Today, electronic systems are embedded in everyday items such as smart phones, mobile computing, biomedical monitoring (bioinstrumentation) systems, entertainment systems and environmental monitoring systems. In many cases, these systems are based on capturing sensor signals, processing and converting them to a suitable digital representation, undertaking digital signal processing (DSP) operations, storing values in local memory, interfacing to a user and finally providing wired or wireless communications to another electronic system. The basic idea is shown in **Figure 1**. The sensors can provide either analogue outputs (such as voltage, current, frequency and impedance) or digital outputs (logic 0 and 1 levels with associated voltage values). In general, the sensor output signals would require signal conditioning in order to create signal values that are in a suitable form to be captured by a digital processing module. This digital processing module would provide the necessary functions in hardware only or as a mixture of

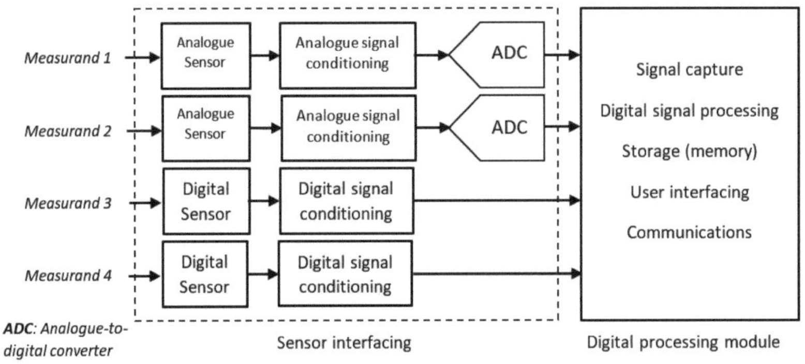

Figure 1. Sensor signal sampling and digital signal processing.

hardware and software operations. The choice of the electronics in the digital processing module in many cases is based on using either software programmable devices such as the microcontroller (*μC*), microprocessor (*μP*) and digital signal processor (*DSP*), or hardware configurable devices such as the field programmable gate array (*FPGA*) and complex programmable logic device (*CPLD*). However, the alternative that involves the design of a custom integrated circuit would be based on application specific integrated circuit (*ASIC*) design techniques. Designing such ASICs would enable a custom design to be created and higher levels of integration that result in physically smaller electronics and the integration of digital, analogue and mixed-signal circuits within a single packaged device. Considering the analogue sensor part of the system, the signal output from the sensor would normally need to be modified (conditioned) in order to provide signal levels that can be sampled by the digital signal-processing module via a suitable ADC, which converts the analogue signal to a digital representation.

Such signal conditioning operations include signal amplification, DC level shifting and anti-aliasing filtering (low-pass filtering to remove any high frequency signal components that would be aliased to lower frequencies). In general, these signal conditioning circuits are based on the use of the op-amp with negative feedback using external resistors and capacitors. However, the operating conditions of the op-amp such as the power supply voltage level would need to be taken into account when either selecting an existing op-amp to use or when designing the op-amp itself. The performance of the op-amp in these types of signal conditioning circuits would be a key factor in what performance could be achieved with the circuits used. In the past, the power supply voltage would not have been a major factor in determining the op-amp performance. The power supply voltage would have been at levels that enabled the op-amp circuitry to operate without encountering power supply voltage limitation issues. With the move towards lower power supply voltage levels at, and below 3.3 V operation, and moving towards 1 V system operation, the power supply conditions must now be accounted for. The op-amp circuit architectures along with circuit design approaches must be reconsidered in order to enable these op-amps to be designed with appropriate characteristics for low-voltage operation.

3. Conventional op-amp design approach

3.1. Introduction

The op-amp is a high-gain DC differential amplifier that is the core building block for many analogue circuits. In general, it consists of two or more amplification stages using transistors, integrated capacitors and in some designs, integrated resistors. **Figure 2** identifies the basic symbol for the voltage input/voltage output op-amp, which has two inputs (the inverting (*IN−*) and non-inverting (*IN+*) inputs), a DC power supply (*V +* and *V−*) and either one output (a single-ended output (a)) or two outputs (a differential output (b)). The op-amp is designed to have certain characteristics that include a high open-loop differential gain (A_{OL}), a high gain-bandwidth product, a high input resistance, a low output resistance, a low output offset voltage, a high dynamic range (minimum to maximum signal range) and a high common-mode rejection ratio (*CMRR*) [1]. The op-amps shown in **Figure 2** identify the circuits in open loop without any

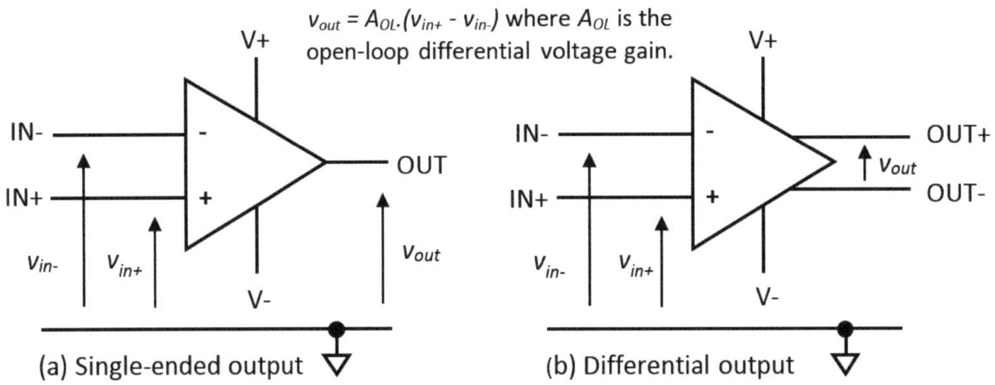

Figure 2. Single-ended output and differential output op-amps.

external feedback components from the output signal back to the input signal. The op-amp, therefore, would have a set of open-loop characteristics. In general, the op-amp would be designed to operate in closed loop where feedback components, primarily resistors and capacitors are used to provide either negative (linear operations) or positive (non-linear operations) feedback.

In the discussion within this chapter, CMOS fabrication process is considered as it is the most widely used fabrication process to realise VLSI ICs. The work presented here will focus on CMOS op-amp circuit design considerations, particularly the AC (frequency) response and stability. The standard topology for the single-ended output two-stage op-amp is considered, and the behaviour of an example case study design will be presented.

3.2. Metal oxide semiconductor field effect transistor

The metal oxide semiconductor field effect transistor (MOSFET) is the most widely used semiconductor device. It is a non-linear device that has four terminals: the drain, source, gate and bulk (or body, substrate). Two forms of MOSFET can be created: the n-channel (nMOS) and p-channel (pMOS) [2]. With these transistors, a voltage between the gate and the source (v_{gs}) controls the flow of drain current (i_d). To design circuits using these devices, it is necessary to know their current-voltage (*IV*) characteristics. In conventional circuit design, the transistor is usually modelled using two discrete models to mathematically describe the *IV* characteristics: a large-signal and a small-signal model. Each model would be used for different design and analysis purposes.

3.3. Large-signal model

A curve that describes the large-signal *IV* characteristic is shown in **Figure 3**. The operation of the transistor is modelled using three different regions according to the values of the gate-source voltage (v_{GS}) and the drain-source voltage (v_{DS}). This models the MOSFET drain current (i_D) against v_{DS} with different values of v_{GS}.

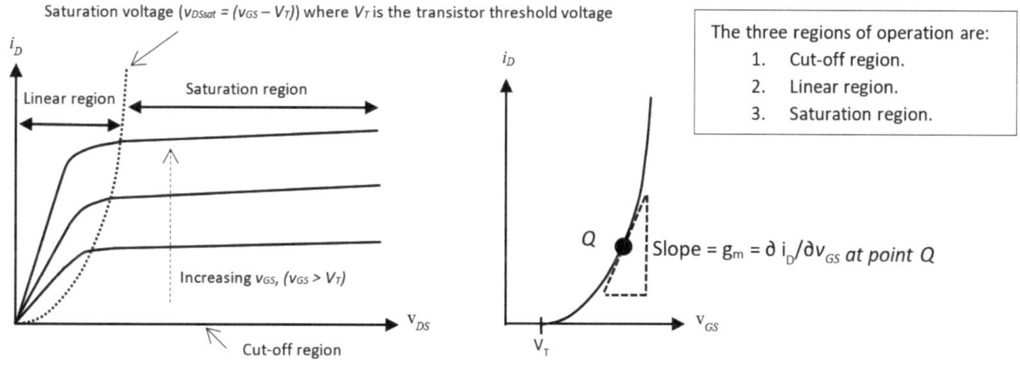

Figure 3. Large-signal *IV* characteristic of the MOSFET.

The three defined regions of operation are *cut-off, linear* and *saturation* where:

Cut-off region: Cut-off is a region in which the transistor will be OFF, and there will be no current flow from the drain to the source (i_D (*cut-off*) = 0). The gate-source voltage is less than the transistor threshold voltage (V_T) in this region.

Linear or ohmic or non-saturation region: In this region, the gate-source voltage is larger than, or equal to, V_T and the drain-source voltage larger than zero but less than the saturation (pinch-off) voltage (v_{DSsat} = (v_{GS}–V_T)). A channel is created between the drain and source terminals, and there is current flow from drain to source. The drain current will increase linearly with increasing drain-source voltage.

Saturation region: In this region, the gate-source voltage is larger or equal to, the transistor threshold voltage, and drain-source voltage has reached or exceeds, v_{DSsat}. This occurs when the channel charge becomes pinched off at the drain-channel interface, and the transistor operation is now in the saturation region. In the simplest (first order) transistor model, increases in v_{DS} do not cause an increase in i_D and so i_D becomes independent of v_{DS}. However, a more representative model includes an i_D dependence on the value of v_{DS}. Moreover, the transistor operation depends on the gate overdrive voltage (v_{eff} = (v_{GS}–V_T)) with the drain-source channel in strong inversion.

3.4. Small-signal model

Although the transistor is a non-linear device, for circuit analysis purposes when developing linear circuits, a linear model for the transistor operating in the saturation region at a specified DC operating (bias) point is initially created. This then describes the behaviour of the transistor to small-signal changes around the bias point, and the small-signal model is then used to determine AC gain values. The signal changes are considered to be small so enabling the approximation that the transistor operation is linear around this DC operating point to be valid. Moreover, defining the small-signal behaviour of the transistor as a transfer function, the transconductance (g_m), and output conductance (g_o) is required model parameters. The small-signal equivalent circuit model for the MOSFET is shown in **Figure 4**.

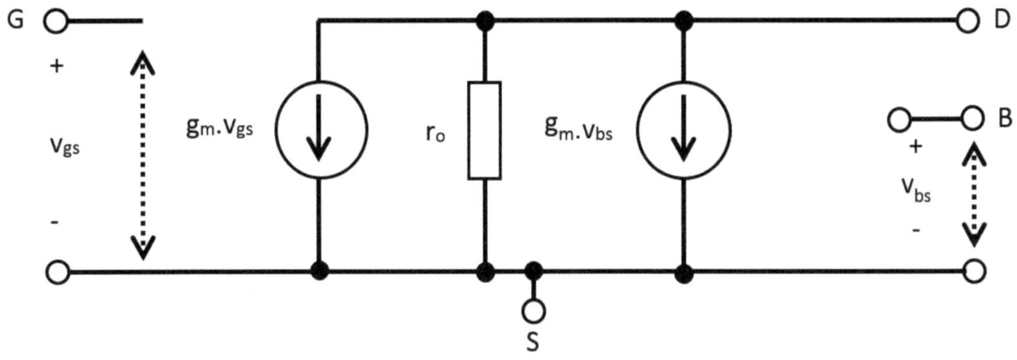

Figure 4. MOSFET small-signal equivalent circuit model.

However, if the signal level is increased, the transistor operation becomes non-linear and will represent by the large-signal model. The conventional analogue design method for the op-amp considers the use of the transistor operating in the saturation region and the drain-source channel to be in strong inversion. This requires the circuit voltage levels (and hence the power supply voltage) to be of suitably high levels to ensure that the transistor remains in saturation and strong inversion for linear circuit operation. Analogue CMOS integrated circuit design needs to use a suitable technology to determine MOSFET dimensions and create the required circuit performance. However, today, when developing circuit designs based on using MOSFETs at low-power and low-voltage, the small-signal and large-signal models are no longer suitable to define transistor operation.

3.5. Example two-stage CMOS op-amp design

The op-amp circuit can be based on different architectures, and each architecture provides advantages in operation when compared to other architectures. In the design considered in this chapter, the two-stage CMOS operational amplifier is used with a simplified architecture as shown in **Figure 5**. Two amplification stages are used, the first stage providing high voltage gain and the second stage providing additional voltage gain and a large output signal swing. In addition, each stage uses negative feedback frequency compensation to improve stability and bandwidth. Negative Miller compensation is applied around the first stage using two identical capacitors (C_{NM}), and Miller compensation is applied around the second stage using two identical capacitors (C_M). The circuit schematic of the selected op-amp architecture is shown in **Figure 6**. Note how the signals between the first stage and the second stage are connected and how the actual circuit connections differ from the simplified architecture (**Figure 5**). The first stage consists of a transconductance stage with differential input transistors *PM1* and *PM2* followed by folded cascode (*FC*) stage. The mirror connected transistors *NM5* and *NM6* in the folded cascode sum the input transistors differential current. The current sources *PM8* and *PM9* on the upper side must provide a current larger than the bias current for each input transistor.

The second stage is a class AB amplifier, and the single-ended output comes from transistors *PM17* and *NM14*. The second stage is primarily used to provide a large output voltage swing

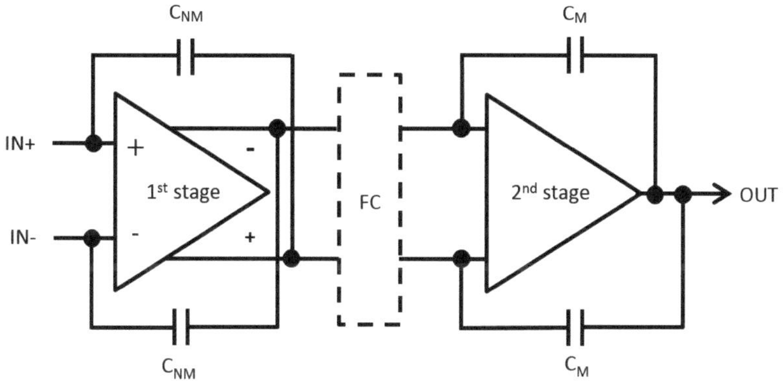

Figure 5. Two-stage op-amp case study design simplified architecture.

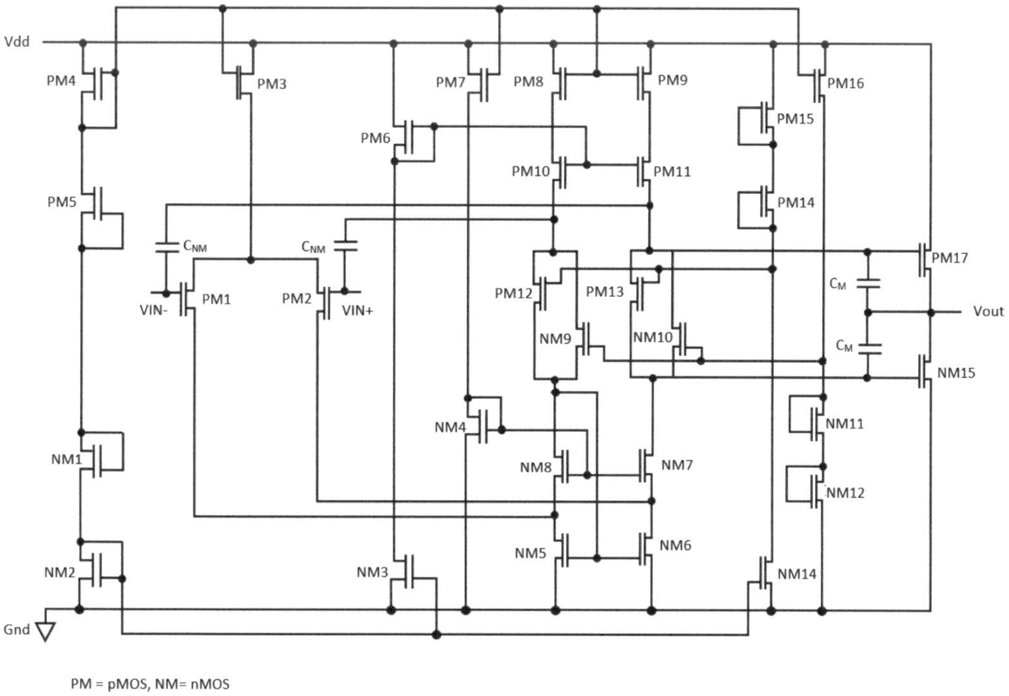

Figure 6. Two-stage op-amp design case study design schematic.

(rail-to-rail output) with high DC voltage gain. *NM10* and *PM13* perform the feed-forward class-AB control. These transistors are biased by two in-phase signal currents using the two cascode transistors *NM8* and *PM11*. The gate voltages for these two transistors are kept at a constant value using the stacked diode-connected transistors (*PM14, PM15* and *NM11, NM12*). The floating current source (*PM12* and *PM13*) has the same structure as the feed-forward class-AB control. The compensation circuitry is split into two parts. Miller compensation around the second stage provides op-amp stability. The op-amp has two Miller capacitors around the class-AB amplifier. Negative Miller compensation around the first stage is provided the extended the bandwidth (increases the unity gain frequency) and also uses two capacitors.

4. Op-amp stability and compensation techniques

4.1. Introduction

The reason for considering stability in a circuit design is to ensure that the circuit remains stable under the required operating conditions. Instability occurs when the op-amp is configured with negative feedback, and under certain conditions, the negative feedback becomes positive. In the unstable case, the circuit output then oscillates. Stability under any input condition is referred to as unconditionally stable, or absolutely stable [3]. However, if a system is not unconditionally stable, a margin of stability must be built-in to ensure stable operation under the required operating conditions. To achieve stable op-amp operation in closed-loop, the designer can add a capacitance between specific nodes within the op-amp that deliberately reduces the open-loop gain magnitude at higher signal frequencies. This technique, referred to as *compensation*, is implemented by typically bypassing one of the internal op-amp gain stages with a high-pass filter. In the simplest sense, a capacitor is connected between the output and input nodes of a gain stage. The purpose is to decrease the gain magnitude to less than unity at frequencies where instability could occur. A single compensation capacitor implementation is widely used in two-stage op-amp designs. However, there are several other techniques used for the op-amp compensation. Improvements to the op-amp performance using the single capacitor compensation approach include the inclusion of a series resistor, buffer or buffer and series resistor. Other techniques, for example, use multiple feedback capacitors connected to different stages within the circuit. These techniques can be used with the two-stage op-amp. Additional techniques require the inclusion of more than two gain stages and, with decreases in integrated circuit process geometries, op-amps with more than two gain stages have become more common to achieve a sufficiently high open-loop gain. There are two common assumptions in the design of compensation topologies. First, the gain magnitude of the stage is larger than one. Second, the compensation and output load capacitance values are larger than the combined output transistor capacitances for each stage. In addition to the DC gain of the op-amp, there are four parameters of particular interest pertaining to its frequency response. These are the unity-gain bandwidth (UGB), gain-bandwidth product (GBWP), phase margin (PM) and gain margin (GM). UGB specifies the frequency at which the op-amp open-loop differential gain magnitude ($|A_{OL}|$) is unity (i.e. $0\ dB$). GBWP defines the gain-bandwidth product of the op-amp gain magnitude and frequency (f):

$$GBWP = |A_{OL}| * f \tag{1}$$

A potential problem, however, of using a multiple-stage op-amp is for unstable circuit behaviour resulting in an oscillatory output signal due to the capacitances within the op-amp circuit and signal feedback paths that exist. The transistor capacitances and parasitic effects due to layout, along with external components, will contribute to the potential for instability when the op-amp is used in a closed-loop configuration, for example, when the op-amp is used in a unity gain buffer configuration. One common way to predict the closed-loop stability of an amplifier is by determining the PM of the open-loop gain response. The PM must be greater than 0° to prevent negative feedback becoming positive feedback thus creating signal oscillation rather than signal amplification. To determine PM at the unity gain frequency, the difference between the amount of signal phase shift and 180° is determined:

$$PM = (180^0 - |\theta|) \; at \; the \; unity \; gain \; frequency \quad (2)$$

where θ is the phase shift of the output signal in degrees (referenced to 0°) when the gain magnitude is unity (0 dB). It is commonly considered that an op-amp in open-loop will require a phase margin of 45° or higher. The GM (in dB) is the difference between the gain magnitude at 180° phase shift and the unity gain magnitude (i.e. 0 dB):

$$GM = (0 \; dB - \; Gain \; magnitude \; (in \; dB)) \; at \; 180^o \; phase \; shift \quad (3)$$

Given the complexity of the input-output relationship of the op-amp, it is common to model the op-amp input-output behaviour in terms of a transfer function for analysis purposes. Typically, a Laplace transfer function is created to model the frequency response and the response is viewed using a Bode plot. The transfer function provides a form for determining important system response characteristics (without solving the complete set of differential equations) in the form:

$$H(s) = \frac{N(s)}{D(s)} = \frac{a_m s^m + a_{m-1} s^{(m-1)} + \ldots + a_2 s + \ldots + a_o}{b_n s^n + b_{n-1} s^{(n-1)} + \ldots + b_2 s + \ldots + b_o} \quad (4)$$

The roots of the numerator *N(s)* (z_x) are called the zeros of the transfer function, and the roots of the denominator *D(s)* (p_y) are called as the poles of the transfer function. *S* is a complex frequency. It is often suitable to factor the polynomials in the numerator and denominator so that the transfer function then becomes:

$$H(s) = \frac{Z(s)}{P(s)} = K.\frac{(s - z_1).(s - z_2)\ldots(s - z_m)}{(s - p_1)(s - p_2)\ldots(s - p_n)} \quad (5)$$

This form of equation directly identifies the system poles and zeros. Using the transfer function characteristics, the Bode plot is a particularly useful tool to visualise the frequency response for analysis purposes. However, with the complexity of the networks formed by the circuit (i.e. the connection of the transistors) and the compensation structures, any system transfer functions that can be derived from the frequency response of the actual circuit to mathematically model the op-amp behaviour rapidly becomes complex. The result is a transfer function with multiple poles and zeros to consider with a complexity that cannot be easy investigated using hand calculations. Hand calculations usually utilise a simplified transfer function, using a form with the most dominant two or three poles, and a full analysis would require the use of a suitable analogue circuit simulator (typically SPICE based) and mathematical modelling tools such as MATLAB. However, deriving simplified transfer function models of the complex circuit can result in loss of detail with some of the critical frequency response parameters. Assumptions are therefore required to simplify the transfer functions without losing important information and any results must be treated with caution, particularly as the relevance of the results obtained must be determined.

4.2. Miller compensation

Miller compensation is achieved by using a capacitor (C_M) between the input and output nodes [4] of the second inverting stage of the two-stage op amp as shown in **Figure 7a**. The dominant (lower frequency) pole in the circuit transfer function is shifted to a lower frequency due to the Miller effect, and the non-dominant (higher frequency) pole is shifted to a higher frequency. The capacitor does not influence the DC response of the amplifier but retains a high gain at mid-band frequencies and reduces the high frequency gain. In this way, the two poles are split and this stabilises the amplifier, but this results in a reduction in signal bandwidth. In addition, the right-hand plane (RHP) zero causes a negative phase shift. The zero comes from the direct feedthrough of the input to the output through the Miller capacitor. If A is the voltage gain of the amplifier, and C_M is a feedback capacitance across the amplifier, Miller theory identifies that C_M effectively shows as a capacitance from the input and output nodes to ground as shown in **Figure 7b**. For a two-stage CMOS op-amp design, considering it to be modelled as a transfer function with two poles only, Miller compensation is used for pole splitting. To establish the frequency dependent gain of this circuit, the small-signal equivalent circuit, as shown in **Figure 8**, can also be created.

The Bode plot for the equivalent circuit in **Figure 8** is shown in **Figure 9** and can be used to identify the positions of the poles and zeros in the transfer function. The first pole ($f_{1'}$) is shifted to a lower frequency (f_1) and the second pole ($f_{2'}$) is shifted to a higher frequency (f_2), although creating the zero (f_z). In addition, the phase is shifted to a higher frequency.

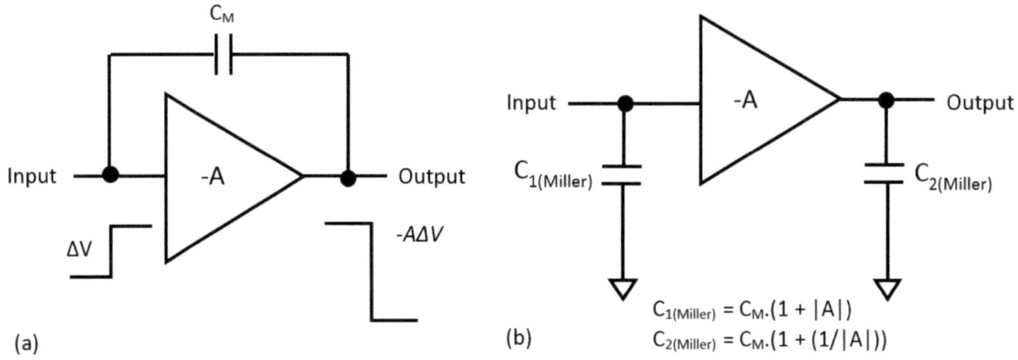

Figure 7. Inverting amplifier with (a) Miller capacitance and (b) equivalent model.

http://dx.doi.org/10.5772/intechopen.68815

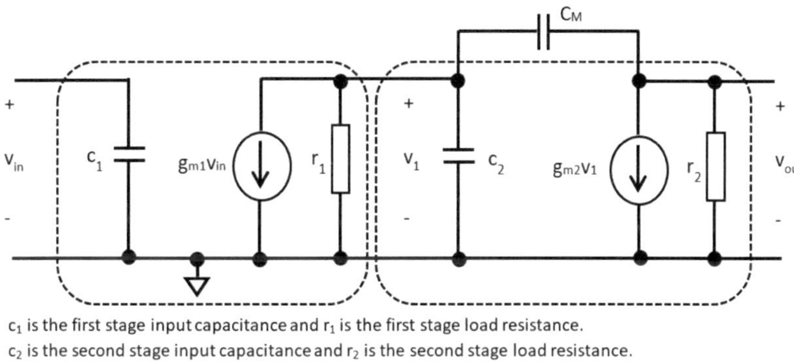

Figure 8. Small-signal equivalent circuit for a two-stage CMOS op-amp including Miller compensation.

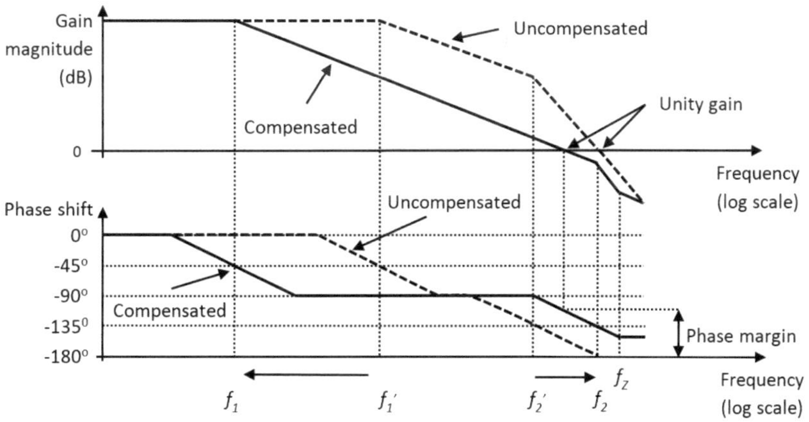

Figure 9. Bode plot showing the pole movement in frequency due to the Miller capacitor.

4.3. Negative Miller compensation

The effect of circuit capacitances, in particular, considering the transistor capacitances, must be considered at higher frequencies as they can cause undesirable phase shifts at higher frequencies that would not be present at lower frequencies. For example, transistor input capacitances can cause problems in circuit operation at higher frequencies and are difficult to eliminate, resulting in reduced op-amp performance. However, as improvement in the fabrication processes leads to reduced transistor geometries, a decrease in transistor capacitance values can be obtained. Negative Miller compensation can, however, be used to improve the frequency response of an op-amp [5]. The idea is shown in **Figure 10**. Negative Miller compensation is based on Miller effect, which defines the effect of the feedback capacitance C_{NM} on the input capacitance C_I. In **Figure 10**, a capacitance (C_{NM}) is connected between the output and input nodes of a non-inverting amplifier. This creates the effect of a negative capacitance. Negative capacitance provides a method for reducing the effects of the transistor input capacitances by the partial cancellation of these capacitances.

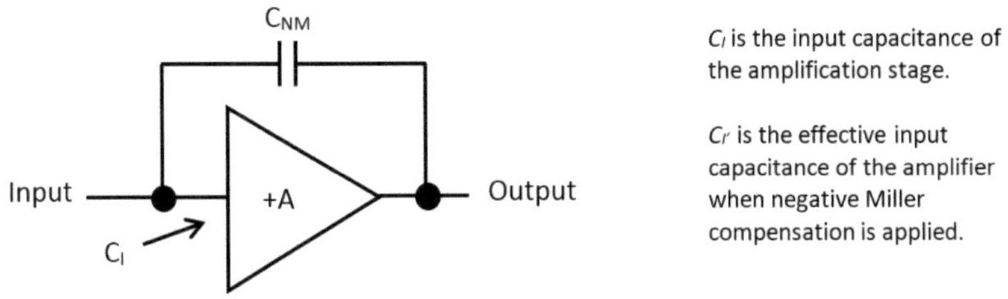

Figure 10. Concept for negative Miller compensation.

The amplifier has a gain magnitude greater than unity. The equivalent input capacitance ($C_I{'}$) is given by:

$$C'_I = C_I + (1 - |A|)C_{NM} \tag{6}$$

The value of C'_I can be controlled by varying C_{NM} and if C_{NM} is large, there will be a net negative capacitance, or an equivalent inductive effect, over a narrow frequency band [6]. Within the op-amp, a gain stage is usually created using an inverting amplifier with a single-ended output. However, with a single-ended output amplifier, in order to incorporate negative Miller compensation then two cascaded gain stages would need to be used. One gain stage would be a non-inverting amplifier, and the feedback capacitor connection is then possible. The second would be an inverting amplifier to provide the overall inverting amplifier arrangement. For a fully differential gain stage, the negative Miller technique can be applied directly. A negative capacitance property can be utilised to improve bandwidth and phase margin. The negative capacitance design moves the non-dominant pole to a higher frequency whilst keeping the location of the dominant pole approximately the same.

5. g_m/I_D ratio design approach

5.1. Introduction

When designing a CMOS op-amp using available transistor models, there can be a substantial difference between the hand calculation results using simple first-order models and simulation results using more complex models (typically BSIM3 transistor simulation models are available for a fabrication process). This would be due to both the complexities of the models used and the accuracy of the models taking into account the boundaries of operation at which the models are designed to operate in. The transistors are, however, operating in the saturation region and in the conventional op-amp design approach, the transistors are considered to also operate in strong inversion where the gate-source voltage is high as discussed in Section 3. At low-voltage operation that is appropriate also for low-power designs, the transistor gate-source voltage is lower and the transistor may be operating in moderate or weak inversion. The transconductance-DC drain current ratio (g_m/I_D) design approach provides separate analytical formulas for strong, moderate and weak inversion, so as to provide simple formulas that are useable in all channel inversion conditions. The approach is particularly suitable for

http://dx.doi.org/10.5772/intechopen.68815

analogue design in CMOS technologies. It considers the relationship between the ratio of the transconductance g_m over DC drain current I_D. In addition, the normalised drain current is also a basic design parameter. The g_m/I_D characteristic provides a useful way to describe the MOSFET operation and provides a straightforward way to estimate transistor dimensions and support circuit design at low-voltage operation. The g_m/I_D ratio is expressed as follows:

$$\frac{g_m}{I_D} = \frac{\partial I_D/\partial V_{gs}}{I_D} = \frac{\partial \log(I_D)}{\partial V_{gs}} \tag{7}$$

Figure 11 identifies two key graphs used. **Figure 11**a on the left shows the g_m/I_D versus V_{GS} characteristic, and **Figure 11**b on the right shows the g_m/I_D versus I_D characteristic. The greater the slope of the curve, the greater the g_m/I_D ratio. This condition occurs when the transistor is operating in weak inversion. As the slope of the curve reduces, the transistor moves into strong inversion. Between weak and strong inversion, moderate inversion occurs. It is to be noted, however, that the region of the moderate inversion is not clearly defined. Weak and moderate inversion are more satisfactory for low-power designs [7]. Moreover, the overdrive voltage (v_{eff}) is low, which is suitable for low supply voltage operation. **Figure 12**a on the left shows the relationship between the g_m/I_D with normalised current $I_D.(W/L)$, and **Figure 12**b on the right shows the transistor transit frequency (f_T) versus g_m/I_D. These curves act as aids to design and hence determining the transistor dimensions. In addition, its analytical form covers all transistor channel inversion conditions, from weak through moderate to strong inversion. The g_m/I_D ratio design approach allows the designer to evaluate design trade-offs for different circuit design operation scenarios.

5.2. MOSFET circuit design from weak to strong inversion

As previously identified, the g_m/I_D ratio is a MOSFET characteristic directly related to all channel inversion conditions [8] of the transistor when the transistor is operating in saturation.

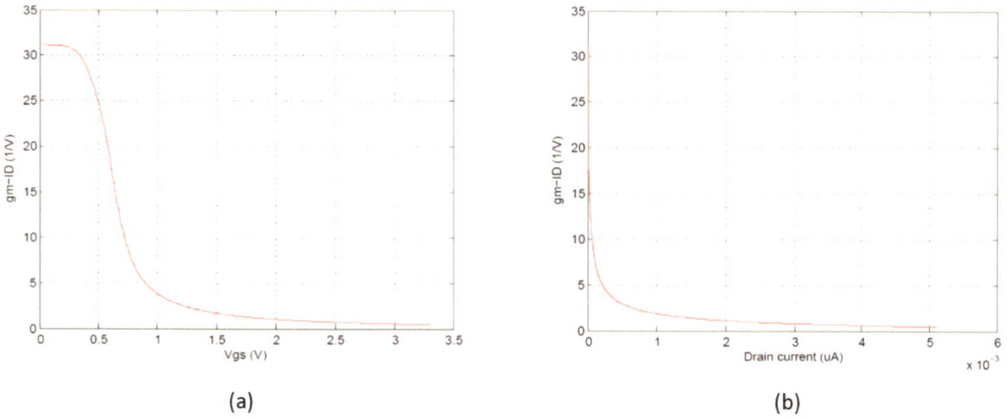

Figure 11. The g_m/I_D ratio of the MOSFET versus: (a) gate-source voltage (V_{GS}) and (b) drain current (I_D).

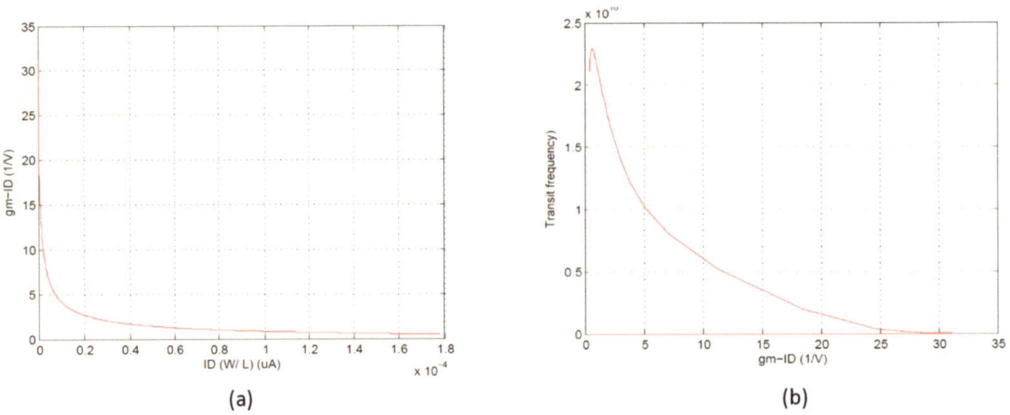

Figure 12. MOSFET characteristics: (a) g_m/I_D ratio versus normalised current ($I_D.(W/L)$) and (b) transit frequency (f_T) versus g_m/I_D ratio.

5.2.1. Strong inversion

When v_{GS} is higher than the threshold voltage V_T, the inversion channel is strongly created, and the drift current is dominant. The classical quadratic i_D-v_{GS} MOSFET equation is based on this condition. The value for g_m in strong inversion is independent of MOSFET sizing and process parameters, and it depends only on the DC bias conditions, I_D and v_{eff}. Similarly, g_m/I_D depends only on v_{eff} and the transistor has a small g_m/I_D, a high gate-source voltage, a high drain current, a high f_T, low noise and small dimensions (width and length).

5.2.2. Moderate inversion

The transition between weak inversion and strong inversion is called moderate inversion. Moderate inversion is important for modern analogue CMOS circuit design where designs are created to operate the MOSFET in this condition. Moderate inversion presents a higher g_m/I_D ratio and a lower gate-source voltage in relation to strong inversion combined with smaller gate area and capacitance, and a higher bandwidth compared to weak inversion.

5.2.3. Weak inversion

In weak inversion, the drain current can be determined using an exponential expression. The transistor has a large g_m/I_D, a low gate-source voltage, a low drain current, a low f_T, high noise and large dimensions (width and length).

Each channel inversion condition has different performance characteristics and a circuit design would then be optimised to account for these characteristics. Given that a design can be created by either using the conventional design approach or the g_m/I_D ratio design approach, **Table 1** provides a summary comparison between the approaches.

http://dx.doi.org/10.5772/intechopen.68815

g_m/I_D design approach	Conventional design approach using V_T, KP and λ
Is valid in all channel operating conditions (weak, moderate and strong) of the MOSFET.	Is valid only in strong inversion of the MOSFET.
Not necessary to create the condition $V_{GS} > V_T$.	Valid only if $V_{GS} > V_T$.
There are different curves for g_m/I_D depending on the inversion region.	Uses the I_D versus V_{GS} and I_D versus V_{DS} curves.
Used for circuits operating on lower power supply voltage levels.	Used for circuits operating on higher power supply voltage levels.
A simplified technique suitable for new evolving fabrication process technologies.	Not suitable for new evolving fabrication process technologies.
A fast design technique as the equations that model the electrical behaviour of circuits can be signified by g_m/I_D.	Not appropriate as a fast design technique. It does not have compact electrical models capable of simple current and voltage relationships.
In this design approach, V_{GS} should be kept as small as possible and transistor gate-source capacitance should be small as possible.	If (V_{GS} $-V_T$) is small, a large geometry device is required and thus large transistor gate-source capacitance.
The g_m/I_D ratio is used directly as a central design variable to determine circuit performance.	The g_m/I_D ratio is not directly used to determine the performance of the circuit.
Links the variables such as g_m, f_T, I_D and V_{eff} to specifications such as bandwidth and power.	Parameters such as μC_{ox}, V_T and $v_{DS(sat)}$ are considered as poorly defined parameters.
Uses charts and simple equations.	Depends on complex equations and sometimes based on assumptions such as ignoring the effect of channel length modulation (λ).
The g_m/I_D ratio associates small-signal and a large-signal ($g_m \rightarrow I_D$) parameters.	Small- and a large-signal models are not associated.

Table 1. Differences between the g_m/I_D ratio design approach and the conventional design approach.

6. Case study op-amp design

6.1. Introduction

The op-amp is an important differential amplifier circuit that has formed the basis of many analogue and mixed-signal IC designs. In this design case study, a two-stage op-amp has been designed and internally compensated by using negative Miller capacitance in the first stage and Miller capacitance in the second stage as shown in **Figure 5**. The idea behind this approach was to develop circuit stability using Miller compensation and increase the bandwidth using negative Miller compensation. The op-amp was designed using the g_m/I_D ratio design approach in order to consider low-voltage operation and is based on the architecture shown in **Figure 5**, with the circuit as shown in **Figure 6**. When operated on a 3.3 V power supply voltage, the MOSFETs operate in moderate inversion to optimise DC gain, unity gain frequency, PM and GM. The op-amp operation was simulated using Cadence Spectre simulator, the MOSFET models were based on a 0.35 μm CMOS fabrication process, and the AC performance both without and with an output load capacitance was assessed in simulation. A differential input voltage was applied to the op-amp in open loop and a single-ended output voltage monitored.

6.2. Op-amp simulation and results

The op-amp simulation study was performed with two conditions: first, no output load capacitance and second, with a variable output load capacitance. An AC analysis was performed on the op-amp design using typical transistor models with the transistors biased for a 3.3 V single-rail power supply voltage operation.

6.2.1. *Study 1: without output load capacitance*

In this study, the simulation approach and results obtained concentrated on the frequency response by using the op-amp with different internal compensation techniques and no output load capacitance. First, no internal compensation was incorporated and then compensation using Miller, negative Miller and a combination of Miller and negative Miller arrangements were considered. **Table 2** shows the results of the simulation study that are shown in Bode plot format in **Figure 13**.

For the op-amp with no compensation and negative Miller compensation only, the GM was a positive number (based on the simulator output value), and hence, the op-amp would be unstable in closed loop. In addition, with these two scenarios, the PM was negative (simulator output value) and this also indicated that the op-amp would be unstable in closed-loop. The results show that the gain magnitude and phase shift are controllable with the different

C_{NM} and C_M = 0.31 pF	No compensation	Negative Miller only	Miller only	Miller and negative Miller
DC gain (dB)	82.5	82.5	82.5	82.5
Phase margin (degrees)	−65.19	−52.56	57.47	63.03
Unity gain frequency (MHz)	1073	1177	177.33	205.54
Gain margin (dB)	14.27	11.08	−10.61	−9.93

Table 2. Open-loop op-amp performance with different compensation techniques.

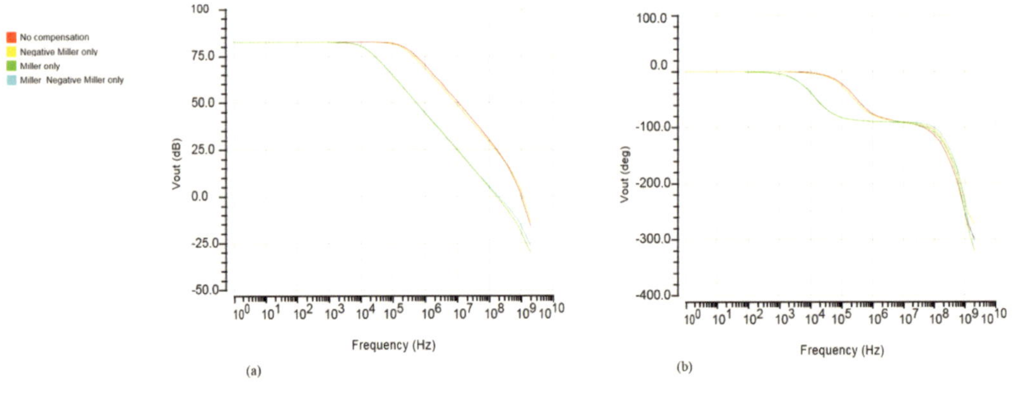

Figure 13. Bode plot of the open-loop op-amp performance with different compensation techniques: (a) gain and (b) phase (Spectre simulation on the transistor circuit model).

compensation techniques and that the choice of compensation technique would determine whether the op-amp is stable or not in closed loop.

6.2.2. Study 2: with output load capacitance

In this study, the simulation approach and results obtained concentrated on the frequency response by using the op-amp with different internal compensation techniques and an output load capacitance with values of 0.1, 0.5 and 1.0 pF. **Table 3** shows the results of the simulation study that are shown in Bode plot format in **Figure 14**.

6.3. Transfer function analysis

An additional form of analysis undertaken with this design was to consider the transfer function for the op-amp input-output relationship. The transfer function is a useful form for evaluating the op-amp frequency response. For a typical op-amp, then the transfer function would contain a large number of poles and zeros. This form would be too complex for initial design development, and so it is common to approximate the transfer function to a simple form that contains typically only two or three poles. These can be estimated from the small-signal equivalent circuit. In addition, once the op-amp design has been created, it is possible to extract the poles and zeros using the circuit simulator and to minimise the initial transfer function of the circuit model having large number of poles and zeros into a simpler transfer

Characteristic	No output load capacitance	0.1 pF	0.5 pF	1.0 pF
DC gain (dB)	82.5	82.5	82.5	82.5
Phase margin (degree)	63.03	61.39	55.99	50.96
Unity gain frequency (MHz)	205.54	203.83	195.54	183.38
Gain margin (dB)	−9.93	−9.80	−9.44	−9.23

Table 3. Open-loop op-amp performance with different output load capacitance values (with combined Miller and negative Miller compensation).

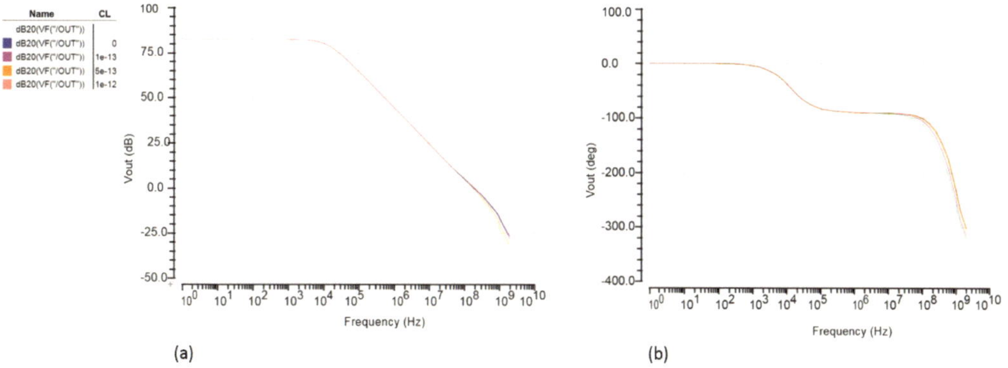

Figure 14. Frequency response of the open-loop op-amp design with different load capacitances: (a) gain and (b) phase (Spectre simulation on the transistor circuit model).

function having a reduced number of poles and zeros. The process for investigating the circuit operation and simplifying the transfer function considered was to:

1. Extract the poles and zeros from the circuit model using Cadence Spectre simulator.
2. Transfer the pole and zero values to MATLAB and create the transfer function.
3. Use MATLAB to reduce the number of poles and zeros in the transfer function and simulate the transfer function behaviour to ensure that the reduced transfer function Bode plot and key characteristics are comparable to the original transfer function.
4. Translate the new transfer function to high-level Verilog-A model and compare the Verilog-A model to the original analogue circuit simulation results and the MATLAB simulation study results.

From the analysis of the open-loop op-amp pole and zero locations as extracted from the circuit, the initial transfer function consisted of 23 poles and 23 zeros. Although it would be expected that the number of poles should be greater than the number of zeros in the transfer function for a strictly proper system, the original 23 pole and zero transfer function extracted is used in the following discussion, and hence, the results are used with a certain level of caution. These were the raw results obtained from the pole-zero analysis in Spectre. It should be noted that the transfer function has the same number of poles and zeros and hence would be referred to as a biproper system. When the transfer function is biproper, it is not reflective of a realisable system at high frequencies as it would have a finite gain at the higher signal frequencies. A strictly proper system where the gain reduces to zero at higher frequencies, as would be expected in a real op-amp, the number of poles must be greater than the number of zeros. This effect can be seen when simulating the transfer function for this design at the higher signal frequencies that would not actually be encountered. To simplify this transfer function from original number of poles and zeros, MATLAB was used to reduce the transfer function to one with just three poles and zeros. **Table 4** shows the resulting performance of the different simulation models, noting that the response of the three models would be valid only up to a certain frequency as the

Number of poles/zeros	Performance	Spectre	MATLAB	Verilog-A
23/23 (MATLAB and Verilog-A transfer function models only)	Gain margin (dB)	9.938	9.95	9.938
	Unity gain frequency (MHz)	205.5	198	205.5
	Phase margin (degrees)	63	63	63
	DC gain (dB)	82.47	82.5	82.47
3/3 (MATLAB and Verilog-A transfer function models only)	Gain margin (dB)	–	10.7	10.64
	Unity gain frequency (MHz)	–	192	200.8
	Phase margin (degrees)	–	54.4	54.61
	DC gain (dB)	–	83.9	83.88

Table 4. Simulated op-amp performance comparison (Spectre (transistor level model), MATLAB (transfer function) and Verilog-A (transfer function)).

transfer functions model a biproper system with a finite high frequency gain rather than a realistic strictly proper transfer function.

7. Op-amp design and operation at lower power supply voltages

Designing and operating analogue circuits at low power supply voltages are challenging tasks. In the past, the circuits typically encountered were designed to operate at higher voltage levels, and so circuit performance limitations due to a limited voltage range was not an issue for many designs. Today, the operation of electronic circuits with low-voltage power supplies is now a requirement for use in electronic systems where size, weight, and power consumption are especially important. For example, in battery-operated portable equipment, a reduction in the battery requirements such as size, weight and energy capacity can provide cost reduction benefits in equipment production, purchase and use as well as making the equipment more portable. The move towards low-voltage operation can be considered from three different perspectives:

1. The increasing use of battery-operated portable systems requires low-power dissipation in order to prolong circuit operation time with a battery energy source.
2. Reduced feature sizes in modern VLSI fabrication processes results in larger electric fields that, unless the power supply voltages are reduced, result in reliability problems.
3. Reduced feature sizes in modern VLSI fabrication processes results in a higher density of the electronics that increases the power dissipation per unit area. The low-voltage operation can be used to reduce the power dissipation per unit area.

As device geometries in CMOS are reduced, the benefits include reduced size, higher operating speeds and reduced power consumption (due to the ability to operate the designs on lower power supply voltage levels), which are mostly exploited in the digital parts of a design. However, this move comes at a cost of introducing device characteristics not seen with larger device geometries. Reducing the power supply voltage has been exploited effectively in digital circuits, but analogue circuits exploiting reduced geometry and voltage operation need to account for a range of circuit performance limiting issues not a concern in digital. In analogue circuits, reducing device geometries and power supply voltage levels have an enormous impact on the analogue circuit capability. For example, as the device geometries become smaller and circuit densities increase, currents in the circuit may need to be reduced in order to prevent excessive temperature increments due to the power consumption per unit area. In addition, reliability problems would exist at higher voltage levels (voltage levels which were commonly used in the past, such as 5 V, but now would be too high for reliable circuit operation) due to excessively high electric fields that would exist. Process variations as CMOS technology move to the lower (deep) sub-micron levels and their effects on low geometry devices, such as transistor width and length dimensions, means that analogue circuit performance can vary widely between devices of the same type and this is accompanied in reduced device geometries by an increase in transistor leakage currents. Whilst the geometries reduce, the transistor threshold voltage (V_T) is, however, remaining relatively constant, and as the power supply voltage is reduced, this causes as reduction in the available voltage range for circuit operation (a reduction in the ($V_{DD} - V_T$) value). Analogue circuits would typically require the creation of bias currents for circuits such as current mirrors which are created using transistors. The need to account for the transistor to be operating in either the weak, moderate or strong inversion regions of operation and the resulting transistor performance differences due to the region of operation would need to be accounted for. The use of fully differential structures is considered given their superior performance with circuit parameters such as *CMRR* and *PSRR* (power-supply rejection-ratio), lower signal distortion and wider signal swing range. Finally, the need to maximise dynamic signal range which often requires a rail-to-rail output voltage range and for op-amps operated in unity-gain configuration, the input stage should also have a rail-to-rail common mode input voltage range [9].

8. Conclusions

The op-amp is an integral part of the on-chip analogue signal conditioning circuitry for the front-end section of mixed-signal ICs. In this chapter, the design of the two-stage op-amp was considered, which was designed using a 0.35 μm CMOS fabrication process and working on a single rail 3.3 V power supply. Considerations were given to low-voltage design (operating at and below 3.3 V) by using the g_m/I_D ratio design approach and the use of both Miller and negative Miller compensation as an internal compensation scheme for op-amp stability and signal bandwidth reasons. The discussion was accompanied by an op-amp case study design and simulation study results that focused on AC performance.

Acknowledgements

The authors would like to acknowledge the support for this work from the Iraqi Ministry of Higher Education and Scientific Research (MOHESR).

Author details

Muhaned Zaidi[1,2], Ian Grout[1]* and Abu Khari A'ain[3]

*Address all correspondence to: ian.grout@ul.ie

1 Department of Electronic and Computer Engineering, University of Limerick, Limerick, Ireland

2 Wasit University, Wasit, Iraq

3 Department of Electronic and Computer Engineering, Universiti Teknologi Malaysia, Skudai, Malaysia

References

[1] Haskard Malcolm R. Analog VLSI design: nMOS and CMOS. Hoboken, NJ, USA: Prentice-Hall, Inc.; 1987

[2] Baker R Jacob. CMOS: Circuit Design, Layout, and Simulation. 3rd ed. Hoboken, NJ, USA: John Wiley & Sons; 2011. DOI: 10.1002/9780470891179

[3] Laker K, Sansen W. Design of Analog Integrated Circuits and Systems. 1st ed. New York, USA: McGraw-Hill Companies; 1994. p. 898

[4] Eduard Säckinger. Broadband Circuits for Optical Fiber Communication. 1st ed. Hoboken, NJ, USA: Wiley; 2005. p. 456

[5] Boaz S-T, Mücahit K, Friedman EG. A high-speed CMOS op-amp design technique using negative Miller capacitance. In: 11th IEEE International Conference on Electronics, Circuits and Systems (ICECS 2004); 15 Dec. 2004; IEEE; 2004. pp. 623-626. DOI: 10.1109/ICECS.2004.1399758

[6] Wu H-M. A 3.125-GHz limiting amplifier for optical receiver system. In: 2006 IEEE Asia Pacific Conference on Circuits and Systems (APCCAS 2006); 4–7 Dec. 2006; IEEE; 2007. pp. 210-213. DOI: 10.1109/APCCAS.2006.342369

[7] Silveira F, Flandre D, Jespers PGA. A gm/ID based methodology for the design of CMOS analog circuits and its application to the synthesis of a silicon-on-insulator micropower OTA. IEEE Journal of Solid-State Circuits. 1996;**31**(9):1314-1319. DOI: 10.1109/4.535416

[8] Jespers P. The gm/ID methodology, a sizing tool for low-voltage analog CMOS circuits. New York, USA: Springer Science & Business Media; 2010. p. 188

[9] Wassenaar RF, Gierkink SLJ, Botma JH. Low-voltage CMOS operational amplifiers. In: Edgar S-S, Andreou AG, editors. Low-Voltage/Low-Power Integrated Circuits and Systems: Low-Voltage Mixed-Signal Circuits (IEEE Press Series on Microelectronic Systems). 1st ed. Piscataway, NJ, USA: Wiley-IEEE Press; 1999

Design of High-Order CMOS Analog Notch Filter with 0.18 μm CMOS Technology

Kittipong Tripetch

Additional information is available at the end of the chapter

http://dx.doi.org/10.5772/intechopen.73157

Abstract

Analog notch filters schematics are very rare. Two circuit diagrams are reviewed with symbolic equations. The first schematic is analog notch filter based on twin-T circuit diagram. The second schematic is analog notch filter based on the Friend biquad circuit.

Keywords: analog notch filter, high-order filter, LCR prototype, interference rejection

1. Introduction

Notch filters or band stop filters have many types of applications. The first application is interference mitigation in GNSS receiver [1]. The second application is the removal of powerline noise from biomedical signals which have operating frequency range from 50 to 60 Hz, while biomedical signal such as EEG has magnitude response in the range of 1–40 Hz [2]. The third application is for a radio frequency image rejection [3]. The fourth application is for an interference rejection in UWB systems. In this application, the filter can notch the magnitude more than 35 db at operating frequency of 900 MHz [4].

A second-order notch can be constructed using an LCR passive prototype. The advent of the very large-scale integration allows tens of thousands of transistors to be fabricated in an integrated circuit. CMOS analog notch filters can be easily designed and built in an IC chip. There are many types of techniques to design analog filter at the architecture or block diagram level such as active RC filter, Gm-C filter, switched Capacitor filter, etc. In this chapter, we will design analog notch filter based on Gm-C filter block diagram.

2. Transconductor capacitor filter based on floating active inductors

There are many choices of transconductor in the literatures. The first transconductor was published by Nedungadi [5]. It is proposed since 1984. This transconductor is very linear; its linear range can be extended by design and simulation. The circuit diagram is shown in **Figure 1**. Its typical linear range, which is output current versus input voltage, can be plotted by level 1 transistor model as follows.

Drain current of an NMOS and a PMOS transistor can be expressed as follows [6]:

$$I_D = \frac{\mu_n C_{ox}}{2}\left(\frac{W}{L}\right)(V_{GS} - V_{TH})^2(1 + \lambda V_{DS}) \tag{1}$$

$$-I_D = \frac{\mu_p C_{ox}}{2}\left(\frac{W}{L}\right)(V_{GS} - V_{TH})^2(1 - \lambda V_{DS}) \tag{2}$$

where I_D is the drain current, μ_n is the electron mobility, μ_p is the hole mobility, C_{ox} is oxide thickness and λ is the channel length modulation.

For submicron CMOS, drain current of NMOS and PMOS transistor can be shown in the formulas (3) and (4). As a consequence of high electric field, both x and y dimensions are a derivative of electric filed by distance along x- and y-axes:

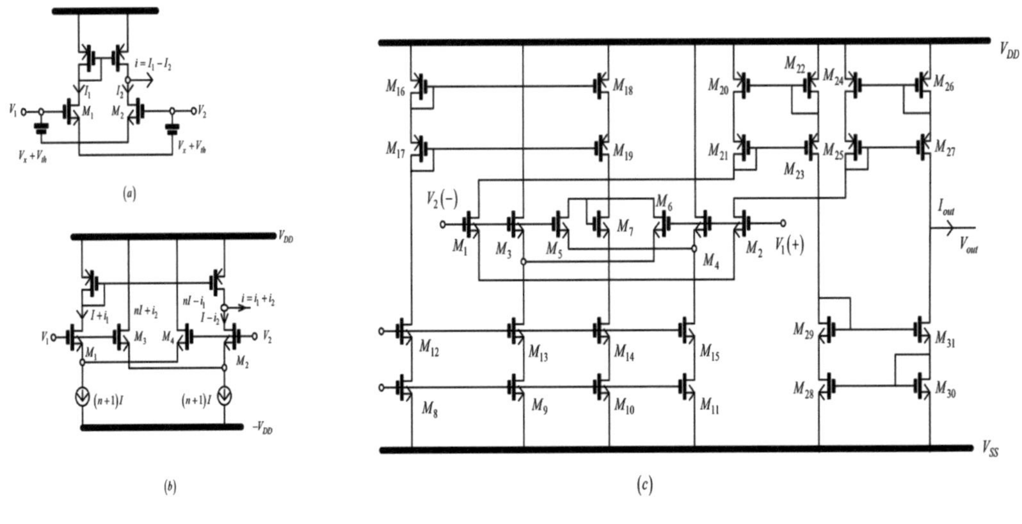

Figure 1. (a) Differential amplifier with cross couple concept, (b) replacement of ideal voltage source with transistor in (a), and (c) cross couple circuit diagram with cascade active load.

$$I_D = \frac{W}{L}\left(\frac{\mu_e C_{ox}}{1+\dfrac{V_{DS}}{E_C L}}\right)\left(V_{GS} - V_{TH} - \frac{V_{DS}}{2}\right)V_{DS}$$

$$\mu_e = \frac{\mu_0}{1+\left(\frac{V_{GS}-V_{TH}}{\theta t_{ox}}\right)^{\eta}}, \eta = 1.85 \quad \textit{for 0.13 } \mu m \tag{3}$$

$$I_{DS} = W v_{sat} C_{ox} \frac{(V_{GS} - V_{TH})^2}{(V_{GS} - V_{TH}) + E_C L} \approx W v_{sat} C_{ox}(V_{GS} - V_{TH})$$

$$E_C L \gg V_{GS} - V_{TH} \quad \textit{for long channel device} \tag{4}$$

$$E_C L \ll V_{GS} - V_{TH} \quad \textit{for short channel device}$$

In order for someone to plot linear range by using multiple transistors, output current can be written as a function input voltage by writing KVL around the loop. Another way of representation is to derive small signal transconductance gain in frequency domain which is a ratio of output current which flows out from the output node divided by input voltage. Small-signal equivalent circuit concept can make the circuit analysis difficult because of parasitic capacitance. Transconductor circuit diagram which has too many transistors may not work if it is believed in small-signal circuit concept because the circuit has too many poles and zeros which make the element substitution of transconductor to deviate from ideal transfer function of LCR prototype.

3. Second-order notch filter

Circuit idea of notch filter is very rare. This is because the theory of an ideal second-order transfer function is well defined. The notch filter or band reject filer is found to be expressed as (5) below [7]:

$$H(s) = \frac{s^2 + \omega_z^2}{s^2 + \left(\frac{\omega_p}{Q_p}\right)s + \omega_p^2} \tag{5}$$

where ω_z is the notch frequency, ω_p is a pole frequency and $\omega_z = \omega_p$.

Numerator polynomial can be designed to have any value so that the roots of the numerator polynomial have roots of it equal with complex zero after equating them with zero.

The circuit which implements this function is called twin-T RC network which can be drawn in **Figures 2** and **3**.

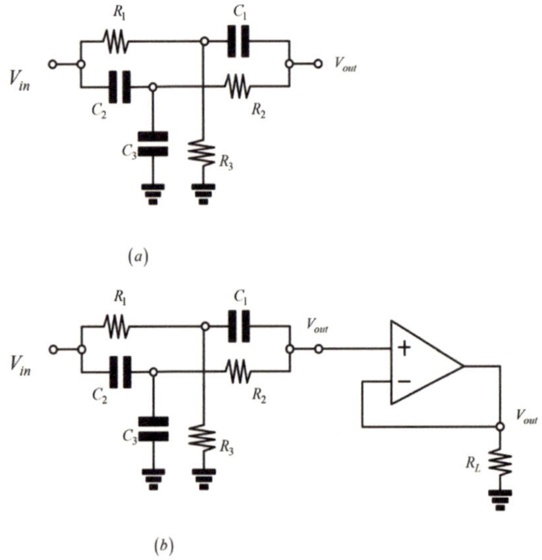

Figure 2. (a) Twin T network and (b) twin T network with buffered op-amp.

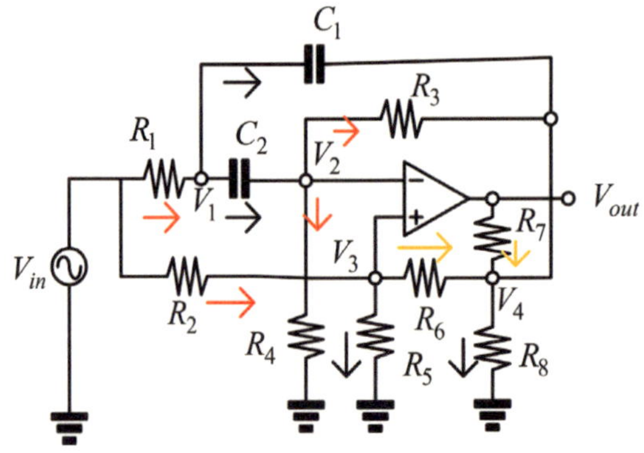

Figure 3. The Friend Biquad circuit.

A. Appendix

The notch filter block diagram is analyzed with Kirchoff current law to prove that it is notch filter transfer function. There are two notch circuits in this appendix. The passive element has its own name without any duplication of names. The current is assumed to flow from left to right and flow from positive potential to ground. Also assume that all nodes in the circuit have positive potential except ground node.

$$\left(\frac{V_{in} - V_1}{R_1}\right) = \frac{V_1}{R_3} + \frac{V_1 - V_{out}}{sC_1} \tag{6}$$

$$\frac{V_{in} - V_2}{sC_2} = \frac{V_2}{sC_3} + \frac{V_2 - V_{out}}{R_2} \tag{7}$$

$$\frac{V_1 - V_{out}}{sC_1} + \frac{V_2 - V_{out}}{R_2} = V_{out}\left(\frac{1}{R_4} + sC_4\right) \tag{8}$$

$$\begin{aligned}
&\left(\frac{V_{in} - V_1}{R_1}\right) = \frac{V_1}{R_3} + \frac{V_1 - V_{out}}{sC_1}\\
&(V_{in} - V_1)(sC_1R_3) = V_1(sC_1R_1) + (V_1 - V_{out})R_1R_3\\
&V_{in}(sC_1R_3) = V_1(sC_1R_1 + sC_1R_3 + R_1R_3) - V_{out}(R_1R_3)
\end{aligned} \tag{9}$$

$$\begin{aligned}
&\frac{V_{in} - V_2}{sC_2} = \frac{V_2}{sC_3} + \frac{V_2 - V_{out}}{R_2}\\
&(V_{in} - V_2)sC_3R_2 = V_2(sC_2R_2) + (V_2 - V_{out})s^2(C_2C_3)\\
&V_{in}(sC_3R_2) = V_2\left(sC_3R_2 + sC_2R_2 + s^2C_2C_3\right) - V_{out}\left(s^2(C_2C_3)\right)\\
&V_2 = \frac{V_{in}(sC_3R_2) + V_{out}\left(s^2(C_2C_3)\right)}{(sC_3R_2 + sC_2R_2 + s^2C_2C_3)}
\end{aligned} \tag{10}$$

$$\begin{aligned}
&\frac{V_1 - V_{out}}{sC_1} + \frac{V_2 - V_{out}}{R_2} = V_{out}\left(\frac{1}{R_4} + sC_4\right) = V_{out}\frac{(1 + sC_4R_4)}{R_4}\\
&(V_1 - V_{out})R_2R_4 + (V_2 - V_{out})sC_1R_4 = V_{out}(1 + sC_4R_4)(sC_1R_2)\\
&V_1(R_2R_4) + V_2(sC_1R_4) = V_{out}\left(sC_1R_2 + s^2C_4R_4C_1R_2 + R_2R_4 + sC_1R_4\right)\\
&V_1 = \left[\frac{V_{out}\left(s^2C_4R_4C_1R_2 + s(C_1R_2 + C_1R_4) + R_2R_4\right) - V_2(sC_1R_4)}{(R_2R_4)}\right]\\
&\quad = \left[\frac{V_{out}\left(s^2a_{21} + sa_{11} + a_{01}\right) - V_2(sC_1R_4)}{(R_2R_4)}\right]
\end{aligned} \tag{11}$$

Substitute Eq. (11) into an Eq. (9):

$$\begin{aligned}
&\left(\frac{V_{in} - V_1}{R_1}\right) = \frac{V_1}{R_3} + \frac{V_1 - V_{out}}{sC_1}\\
&(V_{in} - V_1)(sC_1R_3) = V_1(sC_1R_1) + (V_1 - V_{out})R_1R_3\\
&V_{in}(sC_1R_3) = V_1(sC_1R_1 + sC_1R_3 + R_1R_3) - V_{out}(R_1R_3)\\
&V_{in}(sC_1R_3) = \left[\frac{V_{out}\left(s^2a_{21} + sa_{11} + a_{01}\right) - V_2(sC_1R_4)}{(R_2R_4)}\right](sC_1R_1 + sC_1R_3 + R_1R_3) - V_{out}(R_1R_3)\\
&V_{in}(sC_1R_3(R_2R_4)) = \left[V_{out}\left(s^2a_{21} + sa_{11} + a_{01}\right) - V_2(sC_1R_4)\right](sC_1R_1 + sC_1R_3 + R_1R_3)\\
&\qquad - V_{out}(R_1R_3)(R_2R_4)\\
&V_{in}(sa_{12}) = V_{out}\left[\left(s^2a_{21} + sa_{11} + a_{01}\right)(s(C_1R_1 + sC_1R_3) + R_1R_3) - (R_1R_3)(R_2R_4)\right]\\
&\qquad - V_2(sC_1R_4)(s(C_1R_1 + C_1R_3) + R_1R_3)
\end{aligned} \tag{12}$$

Substitute an Eq. (10) into an Eq. (12):

$$V_{in}(sa_{12}) = V_{out}\left[\left(s^2a_{21}+sa_{11}+a_{01}\right)\left(s(C_1R_1+sC_1R_3)+R_1R_3\right)-(R_1R_3)(R_2R_4)\right]$$
$$-V_2(sC_1R_4)(s(C_1R_1+C_1R_3)+R_1R_3)$$
$$V_{in}(sa_{12}) = V_{out}\left[\left(s^2a_{21}+sa_{11}+a_{01}\right)\left(s(C_1R_1+sC_1R_3)+R_1R_3\right)-(R_1R_3)(R_2R_4)\right]$$
$$-\left[\frac{V_{in}(sC_3R_2)+V_{out}\left(s^2(C_2C_3)\right)}{(sC_3R_2+sC_2R_2+s^2C_2C_3)}\right](sC_1R_4)(s(C_1R_1+C_1R_3)+R_1R_3)$$
$$V_{in}(sa_{12})\begin{pmatrix} s(C_3R_2+C_2R_2) \\ +s^2C_2C_3 \end{pmatrix} = V_{out}\left[\left(s^2a_{21}+sa_{11}+a_{01}\right)\left(s(C_1R_1+sC_1R_3)+R_1R_3\right)-(R_1R_3)(R_2R_4)\right]$$
$$\left(sC_3R_2+sC_2R_2+s^2C_2C_3\right)-\left[V_{in}(sC_3R_2)+V_{out}\left(s^2(C_2C_3)\right)\right](sC_1R_4)(s(C_1R_1+C_1R_3)+R_1R_3) \tag{13}$$

$$V_{in}(sa_{12})\begin{pmatrix} s(C_3R_2+C_2R_2) \\ +s^2C_2C_3 \end{pmatrix} = V_{out}\left[\left(s^2a_{21}+sa_{11}+a_{01}\right)\left(s(C_1R_1+C_1R_3)+R_1R_3\right)-(R_1R_3)(R_2R_4)\right]$$
$$\left(s(C_3R_2+C_2R_2)+s^2C_2C_3\right)-\left[V_{in}(sC_3R_2)+V_{out}\left(s^2(C_2C_3)\right)\right](sC_1R_4)(s(C_1R_1+C_1R_3)+R_1R_3)$$
$$V_{in}\left[s^3(a_{12}C_2C_3)+s^2a_{12}(C_3R_2+C_2R_2)\right]$$
$$= V_{out}\begin{bmatrix} s^3a_{21}(C_1R_1+sC_1R_3)+s^2\begin{pmatrix} a_{21}(R_1R_3-(R_1R_3)(R_2R_4)) \\ +a_{11}(C_1R_1+C_1R_3) \end{pmatrix} \\ +s[(a_{11})(R_1R_3-(R_1R_3)(R_2R_4))+a_{01}(C_1R_1+C_1R_3)] \\ +a_{01}(R_1R_3-(R_1R_3)(R_2R_4)) \end{bmatrix}\left(s(C_3R_2+C_2R_2)+s^2C_2C_3\right) \tag{14}$$
$$-V_{in}\left[s^3(C_3R_2C_1R_4)(C_1R_1+C_1R_3)+sC_3R_2R_1R_3\right]-V_{out}\left[s^3C_2C_3C_1R_4(C_1R_1+C_1R_3)+s^2C_2C_3R_1R_3\right]$$

$$V_{in}\left[s^3(a_{12}C_2C_3)+s^2a_{12}(C_3R_2+C_2R_2)\right] = V_{out}\left[s^3a_{33}+s^2a_{23}+sa_{13}+a_{03}\right]\left(s(C_3R_2+C_2R_2)+s^2C_2C_3\right)$$
$$-V_{in}\left[s^3(C_3R_2C_1R_4)(C_1R_1+C_1R_3)+sC_3R_2R_1R_3\right]-V_{out}\left[s^3C_2C_3C_1R_4(C_1R_1+C_1R_3)+s^2C_2C_3R_1R_3\right]$$
$$a_{33}=a_{21}(C_1R_1+sC_1R_3),\ a_{23}=\begin{pmatrix} a_{21}(R_1R_3-(R_1R_3)(R_2R_4)) \\ +a_{11}(C_1R_1+C_1R_3) \end{pmatrix},$$
$$a_{13}=[(a_{11})(R_1R_3-(R_1R_3)(R_2R_4))+a_{01}(C_1R_1+C_1R_3)]$$
$$a_{03}=a_{01}(R_1R_3-(R_1R_3)(R_2R_4))$$
$$V_{in}\left[s^3(a_{12}C_2C_3+(C_3R_2C_1R_4)(C_1R_1+C_1R_3))+s^2a_{12}(C_3R_2+C_2R_2)+sC_3R_2R_1R_3\right]$$
$$= V_{out}\left[s^3a_{33}+s^2a_{23}+sa_{13}+a_{03}\right]\left(s(C_3R_2+C_2R_2)+s^2C_2C_3\right)$$
$$-V_{out}\left[s^3C_2C_3C_1R_4(C_1R_1+C_1R_3)+s^2C_2C_3R_1R_3\right]$$
$$a_{34}=(a_{12}C_2C_3+(C_3R_2C_1R_4)(C_1R_1+C_1R_3)),\ a_{24}=a_{12}(C_3R_2+C_2R_2),\ a_{14}=C_3R_2R_1R_3 \tag{15}$$

http://dx.doi.org/10.5772/intechopen.73157

$$V_{in}\left[s^3a_{34}+s^2a_{24}+sa_{14}\right]=V_{out}\left[s^3a_{33}+s^2a_{23}+sa_{13}+a_{03}\right]\left(s(C_3R_2+C_2R_2)+s^2C_2C_3\right)-V_{out}\left[s^3a_{35}+s^2a_{25}\right]$$

$$a_{34}=(a_{12}C_2C_3+(C_3R_2C_1R_4)(C_1R_1+C_1R_3)),\ a_{24}=a_{12}(C_3R_2+C_2R_2),\ a_{14}=C_3R_2R_1R_3$$

$$a_{35}=C_2C_3C_1R_4(C_1R_1+C_1R_3),\ a_{25}=C_2C_3R_1R_3$$

$$V_{in}\left[s^3a_{34}+s^2a_{24}+sa_{14}\right]=V_{out}\begin{bmatrix}s^5a_{33}C_2C_3+s^4(a_{23}C_2C_3+a_{33}(C_3R_2+C_2R_2))\\+s^3(a_{23}(C_3R_2+C_2R_2)+a_{13}(C_3R_2+C_2R_2)-a_{35})\\+s^2(a_{13}(C_3R_2+C_2R_2)+a_{03}C_2C_3-a_{25})\\sa_{03}(C_3R_2+C_2R_2)\end{bmatrix}$$

$$\frac{V_{out}}{V_{in}}=\frac{\left[s^3a_{34}+s^2a_{24}+sa_{14}\right]}{\begin{bmatrix}s^5a_{33}C_2C_3+s^4(a_{23}C_2C_3+a_{33}(C_3R_2+C_2R_2))\\+s^3(a_{23}(C_3R_2+C_2R_2)+a_{13}(C_3R_2+C_2R_2)-a_{35})\\+s^2(a_{13}(C_3R_2+C_2R_2)+a_{03}C_2C_3-a_{25})\\sa_{03}(C_3R_2+C_2R_2)\end{bmatrix}}$$

$$=\frac{s\left[s^2a_{34}+sa_{24}+a_{14}\right]}{s\begin{bmatrix}s^4a_{33}C_2C_3+s^3(a_{23}C_2C_3+a_{33}(C_3R_2+C_2R_2))\\+s^2(a_{23}(C_3R_2+C_2R_2)+a_{13}(C_3R_2+C_2R_2)-a_{35})\\+s(a_{13}(C_3R_2+C_2R_2)+a_{03}C_2C_3-a_{25})\\+a_{03}(C_3R_2+C_2R_2)\end{bmatrix}} \tag{16}$$

KCL at V_1:

$$\begin{aligned}&\left(\frac{V_{in}-V_1}{R_1}\right)=(V_1-V_4)sC_1+(V_1-V_2)sC_2\\&V_{in}-V_1=V_1(sC_1R_1+sC_2R_1)-V_2(sC_2R_1)-V_4(sC_1R_1)\\&V_{in}=V_1(sC_1R_1+sC_2R_1+1)-V_2(sC_2R_1)-V_4(sC_1R_1)\\&V_{in}-V_1x_1+V_2x_2+V_4x_3=0\\&x_1=(sC_1R_1+sC_2R_1+1)\\&x_2=(sC_2R_1)\\&x_3=(sC_1R_1)\end{aligned} \tag{17}$$

KCL at V_2:

$$\begin{aligned}
&(V_1 - V_2)sC_2 = \left(\frac{V_2 - V_4}{R_3}\right) + \frac{V_2}{R_4} \\
&V_1(sC_2) = V_2\left(sC_2 + \frac{1}{R_3} + \frac{1}{R_4}\right) - V_4\left(\frac{1}{R_3}\right) \\
&V_1(sC_2) - V_2x_4 + V_4x_5 = 0 \\
&x_4 = \left(sC_2 + \frac{1}{R_3} + \frac{1}{R_4}\right) \\
&x_5 = \left(\frac{1}{R_3}\right)
\end{aligned} \tag{18}$$

KCL at V_3:

$$\begin{aligned}
&\left(\frac{V_{in} - V_3}{R_2}\right) = \frac{V_3}{R_5} + \left(\frac{V_3 - V_4}{R_6}\right) \\
&\frac{V_{in}}{R_2} = V_3\left(\frac{1}{R_2} + \frac{1}{R_5} + \frac{1}{R_6}\right) - V_4\left(\frac{1}{R_6}\right) \\
&V_{in}x_6 - V_3x_7 + V_4x_8 = 0 \\
&x_6 = \frac{1}{R_2} \\
&x_7 = \left(\frac{1}{R_2} + \frac{1}{R_5} + \frac{1}{R_6}\right) \\
&x_8 = \left(\frac{1}{R_6}\right)
\end{aligned} \tag{19}$$

KCL at V_4:

$$\begin{aligned}
&\frac{V_3 - V_4}{R_6} + \left(\frac{V_{out} - V_4}{R_7}\right) = \frac{V_4}{R_8} \\
&\frac{V_3}{R_6} - V_4\left(\frac{1}{R_6} + \frac{1}{R_7} + \frac{1}{R_8}\right) + \frac{V_{out}}{R_7} = 0 \\
&V_3x_9 - V_4x_{10} + V_{out}x_{11} = 0 \\
&x_9 = \frac{1}{R_6} \\
&x_{10} = \left(\frac{1}{R_6} + \frac{1}{R_7} + \frac{1}{R_8}\right) \\
&x_{11} = \frac{1}{R_7}
\end{aligned} \tag{20}$$

KCL at V_{out}:

$$\begin{aligned}
&A_v(V_3 - V_2) = \frac{V_{out} - V_4}{R_7} \\
&-V_2A_v + V_3A_v + \frac{V_4}{R_7} - \frac{V_{out}}{R_7} = 0 \\
&-V_2A_v + V_3A_v + x_{12}V_4 - x_{12}V_{out} = 0 \\
&x_{12} = \frac{1}{R_7}
\end{aligned} \tag{21}$$

All of these equations can be written in matrix form as follows:

$$\begin{bmatrix} 1 & -x_1 & x_2 & 0 & x_3 & 0 \\ 0 & sC_2 & -x_4 & 0 & x_5 & 0 \\ x_6 & 0 & 0 & -x_7 & x_8 & 0 \\ 0 & 0 & 0 & x_9 & -x_{10} & x_{11} \\ 0 & 0 & -A_v & A_v & x_{12} & x_{12} \\ & & & & & \end{bmatrix} \begin{bmatrix} V_{in} \\ V_1 \\ V_2 \\ V_3 \\ V_4 \\ V_{out} \end{bmatrix} = \begin{bmatrix} 0 \\ 0 \\ 0 \\ 0 \\ 0 \\ 0 \end{bmatrix} \tag{22}$$

From Eq. (17), it can be rewritten as follows:

$$\begin{aligned}
&V_{in} - V_1x_1 + V_2x_2 + V_4x_3 = 0 \\
&V_{in} = V_1x_1 - V_2x_2 - V_4x_3
\end{aligned} \tag{23}$$

Substitute Eq. (23) into Eq. (19); we will get the following equation:

$$\begin{aligned}
&V_{in} = V_1x_1 - V_2x_2 - V_4x_3 \\
&V_{in}x_6 - V_3x_7 + V_4x_8 = 0 \\
&(V_1x_1 - V_2x_2 - V_4x_3)x_6 - V_3x_7 + V_4x_8 = 0 \\
&V_1(x_1x_6) - V_2(x_2x_6) - V_3x_7 + V_4(x_8 - x_3x_6) = 0 \\
&V_1(y_1) - V_2(y_2) - V_3x_7 + V_4(y_3) = 0 \\
&y_1 = (x_1x_6) = \frac{[s(C_1R_1 + C_2R_1) + 1]}{R_2} \\
&y_2 = (x_2x_6) = \frac{(sC_2R_1)}{R_2} \\
&y_3 = (x_8 - x_3x_6) = \frac{1}{R_6} - \frac{(sC_1R_1)}{R_2} = \frac{R_2 - (sC_1R_1R_6)}{R_6R_2}
\end{aligned} \tag{24}$$

All of these equations can be written in matrix form as follows:

$$\begin{bmatrix} 1 & -x_1 & x_2 & 0 & x_3 & 0 \\ 0 & sC_2 & -x_4 & 0 & x_5 & 0 \\ 0 & y_1 & -y_2 & -x_7 & y_3 & 0 \\ 0 & 0 & 0 & x_9 & -x_{10} & x_{11} \\ 0 & 0 & -A_v & A_v & x_{12} & x_{12} \\ 0 & 0 & 0 & 0 & 0 & 0 \end{bmatrix} \begin{bmatrix} V_{in} \\ V_1 \\ V_2 \\ V_3 \\ V_4 \\ V_{out} \end{bmatrix} = \begin{bmatrix} 0 \\ 0 \\ 0 \\ 0 \\ 0 \\ 0 \end{bmatrix} \tag{25}$$

From Eq. (24), it can be rewritten as follows:

$$\begin{aligned} & V_1(y_1) - V_2(y_2) - V_3x_7 + V_4(y_3) = 0 \\ & V_1 = \left(\frac{V_2y_2 + V_3x_7 - V_4y_3}{y_1}\right) \end{aligned} \tag{26}$$

Substitute Eq. (26) into Eq. (17); we will get the following equation:

$$\begin{aligned} & V_1 = \left(\frac{V_2y_2 + V_3x_7 - V_4y_3}{y_1}\right) \\ & V_{in} - V_1x_1 + V_2x_2 + V_4x_3 = 0 \\ & V_{in} - \left(\frac{V_2y_2 + V_3x_7 - V_4y_3}{y_1}\right)x_1 + V_2x_2 + V_4x_3 = 0 \\ & V_{in} + V_2\left(x_2 - \frac{y_2x_1}{y_1}\right) - V_3\left(\frac{x_7x_1}{y_1}\right) + V_4\left(x_3 - \frac{y_3x_1}{y_1}\right) = 0 \\ & V_{in} + V_2y_4 - V_3y_5 + V_4y_6 = 0 \\ & y_4 = \left(x_2 - \frac{y_2x_1}{y_1}\right) = sC_2R_1 - \left(\frac{sC_2R_1}{R_2}\right)\left(\frac{s(C_1R_1 + C_2R_1) + 1}{s(C_1R_1 + C_2R_1) + 1}\right)(R_2) = 0 \\ & y_5 = \left(\frac{x_7x_1}{y_1}\right) = \left(\frac{1}{R_2} + \frac{1}{R_5} + \frac{1}{R_6}\right)\left[\frac{s(C_1R_1 + C_2R_1) + 1}{s(C_1R_1 + C_2R_1) + 1}\right]R_2 = \left(\frac{R_2}{R_2} + \frac{R_2}{R_5} + \frac{R_2}{R_6}\right) \\ & y_6 = \left(x_3 - \frac{y_3x_1}{y_1}\right) = sC_1R_1 - \left[\frac{R_2 - sC_1R_1R_6}{R_6R_2}\right]\left[\frac{(s(C_1R_1 + C_2R_1) + 1)R_2}{(s(C_1R_1 + C_2R_1) + 1)}\right] \\ & = sC_1R_1 - \left[\frac{R_2 - sC_1R_1R_6}{R_6}\right] = s(2C_1R_1) - \left(\frac{R_2}{R_6}\right) \end{aligned} \tag{27}$$

All of these equations can be written in matrix form as follows:

$$\begin{bmatrix} 1 & 0 & y_4 & -y_5 & y_6 & 0 \\ 0 & sC_2 & -x_4 & 0 & x_5 & 0 \\ 0 & y_1 & -y_2 & -x_7 & y_3 & 0 \\ 0 & 0 & 0 & x_9 & -x_{10} & x_{11} \\ 0 & 0 & -A_v & A_v & x_{12} & x_{12} \\ 0 & 0 & 0 & 0 & 0 & 0 \end{bmatrix} \begin{bmatrix} V_{in} \\ V_1 \\ V_2 \\ V_3 \\ V_4 \\ V_{out} \end{bmatrix} = \begin{bmatrix} 0 \\ 0 \\ 0 \\ 0 \\ 0 \\ 0 \end{bmatrix} \tag{28}$$

From Eq. (18), it can be rewritten as follows:

$$\begin{aligned} V_1(sC_2) - V_2x_4 + V_4x_5 &= 0 \\ V_1 &= \left(\frac{V_2x_4 - V_4x_5}{sC_2}\right) \end{aligned} \tag{29}$$

Substitute Eq. (29) into Eq. (24); we will get the following equation:

$$\begin{aligned}
&V_1 = \left(\frac{V_2x_4 - V_4x_5}{sC_2}\right) \\
&V_1(y_1) - V_2(y_2) - V_3x_7 + V_4(y_3) = 0 \\
&\left(\frac{V_2x_4 - V_4x_5}{sC_2}\right)(y_1) - V_2(y_2) - V_3x_7 + V_4(y_3) = 0 \\
&V_2\left(\frac{x_4y_1}{sC_2} - y_2\right) - V_3x_7 + V_4\left(y_3 - \frac{x_5y_1}{sC_2}\right) = 0 \\
&V_2y_7 - V_3x_7 + V_4y_8 = 0 \\
&y_7 = \left(\frac{x_4y_1}{sC_2} - y_2\right) = \left(sC_2 + \frac{1}{R_3} + \frac{1}{R_4}\right)\left(\frac{s(C_1R_1 + C_2R_1) + 1}{sC_2R_2}\right) - \left(\frac{sC_2R_1}{R_2}\right) \\
&y_7 = \frac{\left(sC_2 + \frac{1}{R_3} + \frac{1}{R_4}\right)(s(C_1R_1 + C_2R_1) + 1) - (sC_2R_1)sC_2}{sC_2R_2} \\
&= \frac{s^2C_2(C_1R_1) + s\left[C_2 + \left(\frac{1}{R_3} + \frac{1}{R_4}\right)(C_1R_1 + C_2R_1)\right] + \left(\frac{1}{R_3} + \frac{1}{R_4}\right)}{sC_2R_2} \\
&y_8 = \left(y_3 - \frac{x_5y_1}{sC_2}\right) = \frac{R_2 - (sC_1R_1R_6)}{R_6R_2} - \frac{1}{sC_2R_3}\left(\frac{[s(C_1R_1 + C_2R_1) + 1]}{R_2}\right) \\
&= \frac{(R_2 - (sC_1R_1R_6))sC_2R_3 - R_6([s(C_1R_1 + C_2R_1) + 1])}{sC_2R_2R_3R_6} \\
&y_8 = \frac{-s^2(C_1R_1R_6C_2R_3) + s(R_2C_2R_3 - R_6C_1R_1 - R_6C_2R_1) - R_6}{sC_2R_2R_3R_6}
\end{aligned} \tag{30}$$

All of these equations can be written in matrix form as follows:

$$\begin{bmatrix} 1 & 0 & y_4 & -y_5 & y_6 & 0 \\ 0 & sC_2 & -x_4 & 0 & x_5 & 0 \\ 0 & 0 & y_7 & -x_7 & y_8 & 0 \\ 0 & 0 & 0 & x_9 & -x_{10} & x_{11} \\ 0 & 0 & -A_v & A_v & x_{12} & x_{12} \\ 0 & 0 & 0 & 0 & 0 & 0 \end{bmatrix} \begin{bmatrix} V_{in} \\ V_1 \\ V_2 \\ V_3 \\ V_4 \\ V_{out} \end{bmatrix} = \begin{bmatrix} 0 \\ 0 \\ 0 \\ 0 \\ 0 \\ 0 \end{bmatrix} \tag{31}$$

From Eq. (30), it can be rewritten as follows:

$$\begin{aligned} &V_2 y_7 - V_3 x_7 + V_4 y_8 = 0 \\ &V_2 = \left(\frac{V_3 x_7 - V_4 y_8}{y_7}\right) \end{aligned} \tag{32}$$

It is time to eliminate column 3 by Eq. (32) by substituting into Eq. (27):

$$\begin{aligned} &V_2 = \left(\frac{V_3 x_7 - V_4 y_8}{y_7}\right) \\ &V_{in} + V_2 y_4 - V_3 y_5 + V_4 y_6 = 0 \\ &V_{in} + \left(\frac{V_3 x_7 - V_4 y_8}{y_7}\right) y_4 - V_3 y_5 + V_4 y_6 = 0 \\ &V_{in} + V_3\left(\frac{x_7 y_4}{y_7} - y_5\right) + V_4\left(y_6 - \frac{y_8 y_4}{y_7}\right) = 0 \\ &V_{in} + V_3 y_9 + V_4 y_{10} = 0 \\ &y_9 = \left(\frac{x_7 y_4}{y_7} - y_5\right) \\ &y_{10} = \left(y_6 - \frac{y_8 y_4}{y_7}\right) \end{aligned} \tag{33}$$

Update matrix in Eq. (31) by substituting Eq. (33) into as follows:

$$\begin{bmatrix} 1 & 0 & 0 & y_9 & y_{10} & 0 \\ 0 & sC_2 & -x_4 & 0 & x_5 & 0 \\ 0 & 0 & y_7 & -x_7 & y_8 & 0 \\ 0 & 0 & 0 & x_9 & -x_{10} & x_{11} \\ 0 & 0 & -A_v & A_v & x_{12} & x_{12} \\ 0 & 0 & 0 & 0 & 0 & 0 \end{bmatrix} \begin{bmatrix} V_{in} \\ V_1 \\ V_2 \\ V_3 \\ V_4 \\ V_{out} \end{bmatrix} = \begin{bmatrix} 0 \\ 0 \\ 0 \\ 0 \\ 0 \\ 0 \end{bmatrix} \tag{34}$$

It is time to eliminate column 3 by Eq. (32) by substituting into Eq. (29):

$$\begin{aligned} &V_2 = \left(\frac{V_3 x_7 - V_4 y_8}{y_7}\right) \\ &V_1(sC_2) - V_2 x_4 + V_4 x_5 = 0 \\ &V_1(sC_2) - \left(\frac{V_3 x_7 - V_4 y_8}{y_7}\right) x_4 + V_4 x_5 = 0 \\ &V_1(sC_2) - V_3\left(\frac{x_7 x_4}{y_7}\right) + V_4\left(x_5 + \frac{y_8 x_4}{y_7}\right) = 0 \\ &V_1(sC_2) - V_3 y_{11} + V_4 y_{12} = 0 \end{aligned} \tag{35}$$

Update matrix in Eq. (34) by substituting Eq. (35) into as follows:

$$\begin{bmatrix}(1.4)\\(2.3)\\(3.3)\\ \\ \\ \end{bmatrix}\begin{bmatrix}1 & 0 & 0 & y_9 & y_{10} & 0\\ 0 & sC_2 & 0 & -y_{11} & y_{12} & 0\\ 0 & 0 & y_7 & -x_7 & y_8 & 0\\ 0 & 0 & 0 & x_9 & -x_{10} & x_{11}\\ 0 & 0 & -A_v & A_v & x_{12} & x_{12}\\ \end{bmatrix}\begin{bmatrix}V_{in}\\V_1\\V_2\\V_3\\V_4\\V_{out}\end{bmatrix}=\begin{bmatrix}0\\0\\0\\0\\0\\0\end{bmatrix} \tag{36}$$

It is time to eliminate column 3 by Eq. (32) by substituting into Eq. (21):

$$\begin{aligned}
&V_2=\left(\frac{V_3x_7-V_4y_8}{y_7}\right)\\
&-V_2A_v+V_3A_v+x_{12}V_4-x_{12}V_{out}=0\\
&-\left(\frac{V_3x_7-V_4y_8}{y_7}\right)A_v+V_3A_v+x_{12}V_4-x_{12}V_{out}=0\\
&V_3\left(A_v-\frac{x_7A_v}{y_7}\right)+V_4\left(x_{12}+\frac{y_8A_v}{y_7}\right)-V_{out}x_{12}=0\\
&V_3z_1+V_4z_2-V_{out}x_{12}=0
\end{aligned} \tag{37}$$

Update matrix in Eq. (34) by substituting Eq. (37) into as follows:

$$\begin{bmatrix}1 & 0 & 0 & y_9 & y_{10} & 0\\ 0 & sC_2 & 0 & -y_{11} & y_{12} & 0\\ 0 & 0 & y_7 & -x_7 & y_8 & 0\\ 0 & 0 & 0 & x_9 & -x_{10} & x_{11}\\ 0 & 0 & 0 & z_1 & z_2 & -x_{12}\\ 0 & 0 & 0 & 0 & 0 & 0\end{bmatrix}\begin{bmatrix}V_{in}\\V_1\\V_2\\V_3\\V_4\\V_{out}\end{bmatrix}=\begin{bmatrix}0\\0\\0\\0\\0\\0\end{bmatrix} \tag{38}$$

From Eq. (20), it can be rewritten as follows:

$$\begin{aligned}
&V_3x_9-V_4x_{10}+V_{out}x_{11}=0\\
&V_3=\left(\frac{V_4x_{10}-V_{out}x_{11}}{x_9}\right)
\end{aligned} \tag{39}$$

Substitute Eq. (39) into Eq. (37); we will get the following equation:

$$\begin{aligned} &V_3 = \left(\frac{V_4x_{10} - V_{out}x_{11}}{x_9}\right) \\ &V_3z_1 + V_4z_2 - V_{out}x_{12} = 0 \\ &\left(\frac{V_4x_{10} - V_{out}x_{11}}{x_9}\right)z_1 + V_4z_2 - V_{out}x_{12} = 0 \\ &V_4\left(\frac{x_{10}z_1}{x_9} + z_2\right) - V_{out}\left(\frac{x_{11}z_1}{x_9} + x_{12}\right) = 0 \\ &V_4z_3 - V_{out}z_4 = 0 \end{aligned} \tag{40}$$

Update matrix in Eq. (36) by substituting Eq. (40) into as follows:

$$\begin{bmatrix} 1 & 0 & 0 & y_9 & y_{10} & 0 \\ 0 & sC_2 & 0 & -y_{11} & y_{12} & 0 \\ 0 & 0 & y_7 & -x_7 & y_8 & 0 \\ 0 & 0 & 0 & x_9 & -x_{10} & x_{11} \\ 0 & 0 & 0 & 0 & z_3 & -z_4 \\ 0 & 0 & 0 & 0 & 0 & 0 \end{bmatrix} \begin{bmatrix} V_{in} \\ V_1 \\ V_2 \\ V_3 \\ V_4 \\ V_{out} \end{bmatrix} = \begin{bmatrix} 0 \\ 0 \\ 0 \\ 0 \\ 0 \\ 0 \end{bmatrix} \tag{41}$$

Substitute Eq. (39) into Eq. (30); we will get the following equation:

$$\begin{aligned} &V_3 = \left(\frac{V_4x_{10} - V_{out}x_{11}}{x_9}\right) \\ &V_2y_7 - V_3x_7 + V_4y_8 = 0 \\ &V_2y_7 - \left(\frac{V_4x_{10} - V_{out}x_{11}}{x_9}\right)x_7 + V_4y_8 = 0 \\ &V_2y_7 + V_4\left(y_8 - \frac{x_{10}x_7}{x_9}\right) + V_{out}\left(\frac{x_{11}x_7}{x_9}\right) = 0 \\ &V_2y_7 + V_4z_5 + V_{out}z_6 = 0 \end{aligned} \tag{42}$$

Update matrix in Eq. (41) by substituting Eq. (40) into as follows:

$$\begin{bmatrix} 1 & 0 & 0 & y_9 & y_{10} & 0 \\ 0 & sC_2 & 0 & -y_{11} & y_{12} & 0 \\ 0 & 0 & y_7 & 0 & z_5 & z_6 \\ 0 & 0 & 0 & x_9 & -x_{10} & x_{11} \\ 0 & 0 & 0 & 0 & z_3 & -z_4 \\ 0 & 0 & 0 & 0 & 0 & 0 \end{bmatrix} \begin{bmatrix} V_{in} \\ V_1 \\ V_2 \\ V_3 \\ V_4 \\ V_{out} \end{bmatrix} = \begin{bmatrix} 0 \\ 0 \\ 0 \\ 0 \\ 0 \\ 0 \end{bmatrix} \tag{43}$$

Substitute Eq. (39) into Eq. (33); we will get the following equation:

$$\begin{aligned}
&V_3 = \left(\frac{V_4 x_{10} - V_{out} x_{11}}{x_9}\right)\\
&V_{in} + V_3 y_9 + V_4 y_{10} = 0\\
&V_{in} + \left(\frac{V_4 x_{10} - V_{out} x_{11}}{x_9}\right) y_9 + V_4 y_{10} = 0\\
&V_{in} + V_4\left(\frac{x_{10} y_9}{x_9} + y_{10}\right) - V_{out}\left(\frac{x_{11} y_9}{x_9}\right) = 0\\
&V_{in} + V_4 z_7 - V_{out} z_8 = 0
\end{aligned} \tag{44}$$

Update matrix in Eq. (43) by substituting Eq. (44) into as follows:

$$\begin{bmatrix} 1 & 0 & 0 & 0 & z_7 & -z_8 \\ 0 & sC_2 & 0 & -y_{11} & y_{12} & 0 \\ 0 & 0 & y_7 & 0 & z_5 & z_6 \\ 0 & 0 & 0 & x_9 & -x_{10} & x_{11} \\ 0 & 0 & 0 & 0 & z_3 & -z_4 \\ 0 & 0 & 0 & 0 & 0 & 0 \end{bmatrix} \begin{bmatrix} V_{in} \\ V_1 \\ V_2 \\ V_3 \\ V_4 \\ V_{out} \end{bmatrix} = \begin{bmatrix} 0 \\ 0 \\ 0 \\ 0 \\ 0 \\ 0 \end{bmatrix} \tag{45}$$

Substitute Eq. (37) into Eq. (44); we will get the following equation:

$$\begin{aligned}
&V_4 z_3 - V_{out} z_4 = 0\\
&V_4 = V_{out}\left(\frac{z_4}{z_3}\right)\\
&V_{in} + V_4 z_7 - V_{out} z_8 = 0\\
&V_{in} + V_{out}\left(\frac{z_4}{z_3}\right) z_7 - V_{out} z_8 = 0\\
&V_{in} + V_{out}\left(\frac{z_4 z_7}{z_3} - z_8\right) = 0\\
&V_{in} + V_{out} z_9 = 0 \rightarrow \frac{V_{out}}{V_{in}} = -\frac{1}{z_9}
\end{aligned} \tag{46}$$

Author details

Kittipong Tripetch

Address all correspondence to: kan1972@a2.keio.jp

Rajamangala University of Technology Suvarnabhumi, Japan

References

[1] Borio D, Camoriano L, Presti LL. Two-pole and multi pole notch filters: A computationally effective solution for GNSS interference detection and mitigation. IEEE Systems Journal. 2008;**2**(1):38-47

[2] Biswas U, Maniruzzaman Md. Removing power line interference from ECG signal using adaptive filter and notch filter. In: ICEEICT. 2014

[3] Parthasarathy J, Harjani R. Novel Integratable Notch Filter Implementation for 100 dB Image Rejection. I-473-476

[4] Valeese A, Bevilacqua A, Sandner C, Tiebout M, Gerosa A, Neviani A. Analysis and Design of an Integrated Notch Filter for the rejection of interference in UWB systems. IEEE Journal of Solid-State Circuits. February 2009;**44**(2):331-343

[5] Adams WJ, Nedungadi A, Geiger RL. Design of a programmable OTA with multi decade transconductance adjustment. In: ISCAS89. pp. 663-666

[6] Hodges DA, Jackson HG. Analysis and Design of Digital Integrated Circuit. 3rd ed. United States: Mcgraw-Hill; 2004

[7] Daryanani G. Principle of Active Network Synthesis and Design. Singapore: Wiley; 1976